Chaos: A Very Short Introduction

VERY SHORT INTRODUCTIONS are for anyone wanting a stimulating and accessible way into a new subject. They are written by experts, and have been translated into more than 45 different languages.

The series began in 1995, and now covers a wide variety of topics in every discipline. The VSI library now contains over 500 volumes—a Very Short Introduction to everything from Psychology and Philosophy of Science to American History and Relativity—and continues to grow in every subject area.

Titles in the series include the following:

Leonard A. Smith

CHAOS

A Very Short Introduction

OXFORD
UNIVERSITY PRESS

To the memory of Dave Paul Debeer,
A real physicist, a true friend.

Contents

Acknowledgements

This book would not have been possible without my parents, of course, but I owe a greater debt than most to their faith, doubt, and hope, and to the love and patience of a, b, and c. Professionally my greatest debt is to Ed Spiegel, a father of chaos and my thesis Professor, mentor, and friend. I also profited immensely from having the chance to discuss some of these ideas with Jim Berger, Robert Bishop, David Broomhead, Neil Gordon, Julian Hunt, Kevin Judd, Joe Keller, Ed Lorenz, Bob May, Michael Mackey, Tim Palmer, Itamar Procaccia, Colin Sparrow, James Theiler, John Wheeler, and Christine Ziehmann. I am happy to acknowledge discussions with, and the support of, the Master and Fellows of Pembroke College, Oxford. Lastly and largely, I'd like to acknowledge my debt to my students, they know who they are. I am never sure how to react upon overhearing an exchange like: 'Did you know she was Lenny's student?', 'Oh, that explains a lot.' Sorry guys: blame Spiegel.

Preface

The 'chaos' introduced in the following pages reflects phenomena in mathematics and the sciences, systems where (without cheating) small differences in the way things are now have huge consequences in the way things will be in the future. It would be cheating, of course, if things just happened randomly, or if everything continually exploded forever. This book traces out the remarkable richness that follows from three simple constraints, which we'll call *sensitivity*, *determinism*, and *recurrence*. These constraints allow mathematical chaos: behaviour that looks random, but is not random. When allowed a bit of *uncertainty*, presumed to be the active ingredient of forecasting, chaos has reignited a centuries-old debate on the nature of the world.

The book is self-contained, defining these terms as they are encountered. My aim is to show the what, where, and how of chaos; sidestepping any topics of 'why' which require an advanced mathematical background. Luckily, the description of chaos and forecasting lends itself to a visual, geometric understanding; our examination of chaos will take us to the coalface of predictability without equations, revealing open questions of active scientific research into the weather, climate, and other real-world phenomena of interest.

Recent popular interest in the science of chaos has evolved

differently than did the explosion of interest in science a century ago when special relativity hit a popular nerve that was to throb for decades. Why was the public reaction to science's embrace of mathematical chaos different? Perhaps one distinction is that most of us already knew that, sometimes, very small differences can have huge effects. The concept now called 'chaos' has its origins both in science fiction and in science fact. Indeed, these ideas were well grounded in fiction before they were accepted as fact: perhaps the public were already well versed in the implications of chaos, while the scientists remained in denial? Great scientists and mathematicians had sufficient courage and insight to foresee the coming of chaos, but until recently mainstream science required a good solution to be well behaved: fractal objects and chaotic curves were considered not only deviant, but the sign of badly posed questions. For a mathematician, few charges carry more shame than the suggestion that one's professional life has been spent on a badly posed question. Some scientists still dislike problems whose results are expected to be irreproducible even in theory. The solutions that chaos requires have only become widely acceptable in scientific circles recently, and the public enjoyed the 'I told you so' glee usually claimed by the 'experts'. This also suggests why chaos, while widely nurtured in mathematics and the sciences, took root within applied sciences like meteorology and astronomy. The applied sciences are driven by a desire to understand and predict reality, a desire that overcame the niceties of whatever the formal mathematics of the day. This required rare individuals who could span the divide between our models of the world and the world as it is without convoluting the two; who could distinguish the mathematics from the reality and thereby extend the mathematics.

As in all *Very Short Introductions*, restrictions on space require entire research programmes to be glossed over or omitted; I present a few recurring themes in context, rather than a series of shallow descriptions. My apologies to those whose work I have omitted, and my thanks to Luciana O'Flaherty (my editor), Wendy Parker, and Lyn Grove for help in distinguishing between what was most interesting to me and what I might make interesting to the reader.

How to read this introduction

While there is some mathematics in this book, there are no equations more complicated than $X = 2$. Jargon is less easy to discard. Words in **bold italics** you will have to come to grips with; these are terms that are central to chaos, brief definitions of these words can be found in the Glossary at the end of the book. *Italics* is used both for emphasis and to signal jargon needed for the next page or so, but which is unlikely to recur often throughout the book.

Any questions that haunt you would be welcome online at *http://cats.lse.ac.uk/forum/* on the discussion forum VSI Chaos. More information on these terms can be found rapidly at Wikipedia *http://www.wikipedia.org/* and *http://cats.lse.ac.uk/preditcability-wiki/* , and in the Further reading.

List of illustrations

Figures 7, 8, 9, 11, 12, 13, 19, and 20 were produced with the assistance of Hailiang Du. Figures 24 and 30 were produced with the assistance of Reason Machete. Figures 4 and 29 were produced with the assistance of Martin Leutbecher with data kindly made available by the European Centre for Medium-Range Weather Forecasting. Figure 27 is after M. Hume et al., The UKIPO2 Scientific Report, Tyndal Centre, University of East Anglia, Norwich, UK.

The publisher and the author apologize for any errors or omissions in the above list. If contacted they will be pleased to rectify these at the earliest opportunity.

Chapter 1
The emergence of chaos

Embedded in the mud, glistening green and gold and black,
was a butterfly, very beautiful and very dead.
It fell to the floor, an exquisite thing, a small thing
that could upset balances and knock down a line of
small dominoes and then big dominoes and then
gigantic dominoes, all down the years across Time.

Ray Bradbury (1952)

Three hallmarks of mathematical chaos

The 'butterfly effect' has become a popular slogan of chaos. But is it
really so surprising that minor details sometimes have major
impacts? Sometimes the proverbial minor detail is taken to be the
difference between a world with some butterfly and an alternative
universe that is exactly like the first, except that the butterfly is
absent; as a result of this small difference, the worlds soon come to
differ dramatically from one another. The mathematical version of
this concept is known as ***sensitive dependence***. Chaotic systems
not only exhibit sensitive dependence, but two other properties as
well: they are ***deterministic***, and they are ***nonlinear***. In this
chapter, we'll see what these words mean and how these concepts
came into science.

Chaos is important, in part, because it helps us to cope with unstable systems by improving our ability to describe, to understand, perhaps even to forecast them. Indeed, one of the myths of chaos we will debunk is that chaos makes forecasting a useless task. In an alternative but equally popular butterfly story, there is one world where a butterfly flaps its wings and another world where it does not. This small difference means a tornado appears in only one of these two worlds, linking chaos to uncertainty and prediction: in which world are we? Chaos is the name given to the mechanism which allows such rapid growth of uncertainty in our mathematical models. The image of chaos amplifying uncertainty and confounding forecasts will be a recurring theme throughout this Introduction.

Whispers of chaos

Warnings of chaos are everywhere, even in the nursery. The warning that a kingdom could be lost for the want of a nail can be traced back to the 14th century; the following version of the familiar nursery rhyme was published in *Poor Richard's Almanack* in 1758 by Benjamin Franklin:

> For want of a nail the shoe was lost,
> For want of a shoe the horse was lost,
> and for want of a horse the rider was lost,
> being overtaken and slain by the enemy,
> all for the want of a horse-shoe nail.

We do not seek to explain the seed of instability with chaos, but rather to describe the growth of uncertainty *after* the initial seed is sown. In this case, explaining how it came to be that the rider was lost due to a missing nail, not the fact that the nail had gone missing. In fact, of course, there either was a nail or there was not. But Poor Richard tells us that if the nail hadn't been lost, then the kingdom wouldn't have been lost either. We will often explore the properties of chaotic systems by considering the impact of slightly different situations.

The study of chaos is common in applied sciences like astronomy, meteorology, population biology, and economics. Sciences making accurate observations of the world along with quantitative predictions have provided the main players in the development of chaos since the time of Isaac Newton. According to Newton's Laws, the future of the solar system is completely determined by its current state. The 19th-century scientist Pierre Laplace elevated this determinism to a key place in science. A world is deterministic if its current state completely defines its future. In 1820, Laplace conjured up an entity now known as 'Laplace's demon'; in doing so, he linked determinism and the ability to predict in principle to the very notion of success in science.

> We may regard the present state of the universe as the effect of its past and the cause of its future. An intellect which at a certain moment would know all forces that set nature in motion, and all positions of all items of which nature is composed, if this intellect were also vast enough to submit these data to analysis, it would embrace in a single formula the movements of the greatest bodies of the universe and those of the tiniest atom; for such an intellect nothing would be uncertain and the future just like the past would be present before its eyes.

Note that Laplace had the foresight to give his demon three properties: exact knowledge of the Laws of Nature ('all the forces'), the ability to take a snapshot of the exact state of the universe ('all the positions'), and infinite computational resources ('an intellect vast enough to submit these data to analysis'). For Laplace's demon, chaos poses no barrier to prediction. Throughout this Introduction, we will consider the impact of removing one or more of these gifts.

From the time of Newton until the close of the 19th century, most scientists were also meteorologists. Chaos and meteorology are closely linked by the meteorologists' interest in the role uncertainty plays in weather forecasts. Benjamin Franklin's interest in meteorology extended far beyond his famous experiment of flying a kite in a thunderstorm. He is credited with noting the general movement of the weather from west towards the east and testing this theory by writing letters from Philadelphia to cities further east. Although the letters took longer to arrive than the weather, these are arguably early weather forecasts. Laplace himself discovered the law describing the decrease of atmospheric pressure with height. He also made fundamental contributions to the theory of errors: when we make an observation, the measurement is never exact in a mathematical sense, so there is always some uncertainty as to the 'True' value. Scientists often say that any uncertainty in an observation is due to *noise*, without really defining exactly what the noise is, other than that which obscures our vision of whatever we are trying to measure, be it the length of a table, the number of rabbits in a garden, or the midday temperature. Noise gives rise to *observational uncertainty*, chaos helps us to understand how small uncertainties can become large uncertainties, once we have a model for the noise. Some of the insights gleaned from chaos lie in clarifying the role(s) noise plays in the dynamics of uncertainty in the quantitative sciences. Noise has become much more interesting, as the study of chaos forces us to look again at what we might mean by the concept of a 'True' value.

Twenty years after Laplace's book on probability theory appeared, Edgar Allan Poe provided an early reference to what we would now call chaos in the atmosphere. He noted that merely moving our hands would affect the atmosphere all the way around the planet. Poe then went on to echo Laplace, stating that the mathematicians of the Earth could compute the progress of this hand-waving 'impulse', as it spread out and forever altered the state of the atmosphere. Of course, it is up to us whether or not we choose to wave our hands: free will offers another source of seeds that chaos might nurture.

In 1831, between the publication of Laplace's science and Poe's fiction, Captain Robert Fitzroy took the young Charles Darwin on his voyage of discovery. The observations made on this voyage led Darwin to his theory of natural selection. Evolution and chaos have more in common than one might think. First, when it comes to language, both 'evolution' and 'chaos' are used simultaneously to refer both to phenomena to be explained and to the theories that are supposed to do the explaining. This often leads to confusion between the description and the object described (as in 'confusing the map with the territory'). Throughout this Introduction we will see that confusing our mathematical models with the reality they aim to describe muddles the discussion of both. Second, looking more deeply, it may be that some ecosystems evolve as if they were chaotic systems, as it may well be the case that small differences in the environment have immense impacts. And evolution has contributed to the discussion of chaos as well. This chapter's opening quote comes from Ray Bradbury's 'A Sound Like Thunder', in which time-travelling big game hunters accidentally kill a butterfly, and find the future a different place when they return to it. The characters in the story imagine the impact of killing a mouse, its death cascading through generations of lost mice, foxes, and lions, and:

> all manner of insects, vultures, infinite billions of life forms are thrown into chaos and destruction ... Step on a mouse and you leave your print, like a Grand Canyon, across Eternity. Queen Elizabeth might never be born, Washington might not cross the Delaware, there might never be a United States at all. So be careful. Stay on the Path. Never step off!

Needless to say, someone does step off the Path, crushing to death a beautiful little green and black butterfly. We can only consider these 'what if' experiments within the fictions of mathematics or literature, since we have access to only one realization of reality.

The origins of the term 'butterfly effect' are appropriately shrouded in mystery. Bradbury's 1952 story predates a series of scientific papers on chaos published in the early 1960s. The meteorologist Ed Lorenz once invoked sea gulls' wings as the agent of change, although the title of that seminar was not his own. And one of his early computer-generated pictures of a chaotic system does resemble a butterfly. But whatever the incarnation of the 'small difference', whether it be a missing horse shoe nail, a butterfly, a sea gull, or most recently, a mosquito 'squished' by Homer Simpson, the idea that small differences can have huge effects is not new. Although silent regarding the origin of the small difference, chaos provides a description for its rapid amplification to kingdom-shattering proportions, and thus is closely tied to forecasting and predictability.

The first weather forecasts

Like every ship's captain of the time, Fitzroy had a deep interest in the weather. He developed a barometer which was easier to use onboard ship, and it is hard to overestimate the value of a barometer to a captain lacking access to satellite images and radio reports. Major storms are associated with low atmospheric pressure; by providing a quantitative measurement of the pressure, and thus how fast it is changing, a barometer can give life-saving information on what is likely to be over the horizon. Later in life, Fitzroy became the first head of what would become the UK Meteorological Office and exploited the newly deployed telegraph to gather observations and issue summaries of the current state of the weather across Britain. The telegraph allowed weather information to outrun the weather itself for the first time. Working with LeVerrier of France, who became famous for using Newton's Laws to discover two new planets, Fitzroy contributed to the first international efforts at real-time weather forecasting. These forecasts were severely criticized by Darwin's cousin, statistician Francis Galton, who himself published the first weather chart in the *London Times* in 1875, reproduced in Figure 1.

WEATHER CHART, MARCH 31, 1875.

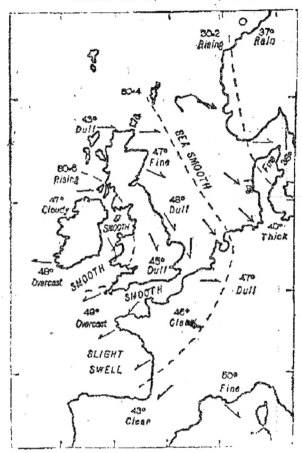

The dotted lines indicate the gradations of barometric pressure. The variations of the temperature are marked by figures, the state of the sea and sky by descriptive words, and the direction of the wind by arrows—barbed and feathered according to its force. ⊙ denotes calm.

1. The first weather chart ever published in a newspaper. Prepared by Francis Galton, it appeared in the *London Times* on 31 March 1875

If uncertainty due to errors of observation provides the seed that chaos nurtures, then understanding such uncertainty can help us better cope with chaos. Like Laplace, Galton was interested in the 'theory of errors' in the widest sense. To illustrate the ubiquitous 'bell-shaped curve' which so often seems to reflect measurement errors, Galton created the 'quincunx', which is now called a Galton Board; the most common version is shown on the left side of Figure 2. By pouring lead shot into the quincunx, Galton simulated a random system in which each piece of shot has a 50:50 chance of going to either side of every 'nail' that it meets, giving rise to a bell-shaped distribution of lead. Note there is more here than the one-off flap of a butterfly wing: the paths of two nearby pieces of lead may stay together or diverge at each level. We shall return to Galton Boards in Chapter 9, but we will use random numbers from the bell-shaped curve as a model for noise many times before then. The bell-shape can be seen at the bottom of the Galton Board on the left of Figure 2, and we will find a smoother version towards the top of Figure 10.

The study of chaos yields new insight into why weather forecasts remain unreliable after almost two centuries. Is it due to our missing minor details in today's weather which then have major impacts on tomorrow's weather? Or is it because our methods, while better than Fitzroy's, remain imperfect? Poe's early atmospheric incarnation of the butterfly effect is complete with the idea that science could, if perfect, predict everything physical. Yet the fact that sensitive dependence would make detailed forecasts of the weather difficult, and perhaps even limit the scope of physics, has been recognized within both science and fiction for some time. In 1874, the physicist James Clerk Maxwell noted that a sense of proportion tended to accompany success in a science:

> This is only true when small variations in the initial circumstances produce only small variations in the final state of the system. In a great many physical phenomena this condition is satisfied; but there are other cases in which a small initial variation may produce a very

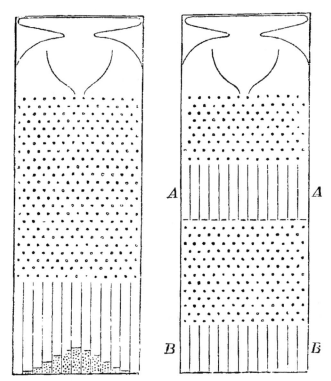

2. Galton's 1889 schematic drawings of what are now called 'Galton Boards'

great change in the final state of the system, as when the displacement of the 'points' causes a railway train to run into another instead of keeping its proper course.

This example is again atypical of chaos in that it is 'one-off' sensitivity, but it does serve to distinguish sensitivity and uncertainty: this sensitivity is no threat as long as there is no uncertainty in the position of the points, or in which train is on which track. Consider pouring a glass of water near a ridge in the

Rocky Mountains. On one side of this continental divide the water finds its way into the Colorado River and to the Pacific Ocean, on the other side the Mississippi River and eventually the Atlantic Ocean. Moving the glass one way or the other illustrates sensitivity: a small change in the position of the glass means a particular molecule of water ends up in a different ocean. Our uncertainty in the position of the glass might restrict our ability to predict which ocean that molecule of water will end up in, but only *if* that uncertainty crosses the line of the continental divide. Of course, *if* we were really trying to do this, we would have to question whether any such mathematical line actually divided continents, as well as the other adventures the molecule of water might have which could prevent it reaching the ocean. Usually, chaos involves much more than a single one-off 'tripping point'; it tends to more closely resemble a water molecule that repeatedly evaporates and falls in a region where there are continental divides all over the place.

Nonlinearity is defined by what it is not (it is not linear). This kind of definition invites confusion: how would one go about defining a biology of non-elephants? The basic idea to hold in mind now is that a nonlinear system will show a disproportionate response: the impact of adding a second straw to a camel's back could be much bigger (or much smaller) than the impact of the first straw. Linear systems always respond proportionately. Nonlinear systems need not, giving nonlinearity a critical role in the origin of sensitive dependence.

The Burns' Day storm

> But Mousie, thou art no thy lane,
> In proving foresight may be vain:
> The best-laid schemes o mice an men
> Gang aft agley,
> An lea'e us nought but grief an pain,
> For promis'd joy!

Still thou art blest, compar'd wi me!
The present only toucheth thee:
But och! I backward cast my e'e,
On prospects drear!
An forward, tho I canna see,
I guess an fear!
  Robert Burns, 'To A Mouse' (1785)

Burns' poem praises the mouse for its ability to live only in the present, not knowing the pain of unfulfilled expectations nor the dread of uncertainty in what is yet to pass. And Burns was writing in the 18th century, when mice and men laid their plans with little assistance from computing machines. While foresight may be pain, meteorologists struggle to foresee tomorrow's likely weather every day. Sometimes it works. In 1990, on the anniversary of Burns' birth, a major storm ripped through northern Europe, including the British Isles, causing significant property damage and loss of life. The centre of the storm passed over Burns' home town in Scotland, and it became known as the Burns' Day storm. A weather chart reflecting the storm at noon on 25 January is shown in the top panel of Figure 4 (page 14). Ninety-seven people died in northern Europe, about half of this number in Britain, making it the highest death toll of any storm in 40 years; about 3 million trees were blown down, and total insurance costs reached £2 billion. Yet the Burns' Day storm has not joined the rogues' gallery of famously failed forecasts: it was well forecast by the Met Office.

In contrast, the Great Storm of 1987 is famous for a BBC television meteorologist's broadcast the night before, telling people *not* to worry about rumours from France that a hurricane was about to strike England. Both storms, in fact, managed gusts of over 100 miles per hour, and the Burns' Day storm caused much greater loss of life; yet 20 years after the event, the Great Storm of 1987 is much more often discussed, perhaps exactly because the Burns' Day storm *was* well forecast. The story leading up to this forecast beautifully illustrates a different way that chaos in our models can impact our lives without invoking alternate worlds, some with and some without butterflies.

In the early morning of 24 January 1990, two ships in the mid-Atlantic sent routine meteorological observations from positions that happened to straddle the centre of what would become the Burns' Day storm. The forecast models run with these observations give a fine forecast of the storm. Running the model again after the event showed that when these observations are omitted, the model predicts a weaker storm in the wrong place. Because the Burns' Day storm struck during the day, the failure to provide forewarning would have had a huge impact on loss of life, so here we have an example where a few observations, had they not been made, would have changed the forecast and hence the course of human events. Of course, an ocean weather ship is harder to misplace than a horse shoe nail. There is more to this story, and to see its relevance we need to look into how weather models 'work'.

Operational weather forecasting is a remarkable phenomenon in and of itself. Every day, observations are taken in the most remote locations possible, and then communicated and shared among national meteorological offices around the globe. Many different nations use this data to run their computer models. Sometimes an observation is subject to plain old mistakes, like putting the temperature in the box for wind speed, or a typo, or a glitch in transition. To keep these mistakes from corrupting the forecast, incoming observations are subject to quality control: observations that disagree with what the model is expecting (given its last forecast) can be rejected, especially if there are no independent, nearby observations to lend support to them. It is a well-laid plan. Of course, there are rarely any 'nearby' observations of any sort in the middle of the Atlantic, and the ship observations showed the development of a storm that the model had not predicted would be there, so the computer's automatic quality control program simply rejected these observations.

3. Headline from *The Times* the day after the Burns' Day storm

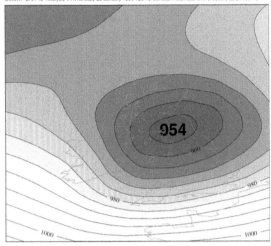

FC +48 h

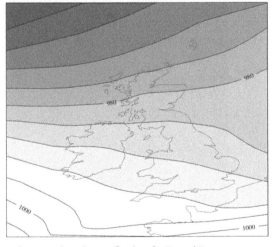

4. A modern weather chart reflecting the Burns' Day storm as seen through a weather model (top) and a two-day-ahead forecast targeting the same time showing a fairly pleasant day (bottom)

Luckily, the computer was overruled. An intervention forecaster was on duty and realized that these observations were of great value. His job was to intervene when the computer did something obviously silly, as computers are prone to do. In this case, he tricked the computer into accepting the observations. Whether or not to take this action is a judgement call: there was no way to know at the time which action would yield a better forecast. The computer was 'tricked', the observation was used. The storm was forecast, and lives were saved.

There are two take-home messages here: the first is that when our models are chaotic then small changes in our observations can have large impacts on the quality of our foresight. An accountant looking to reduce costs and computing the typical benefit of one particular observation from any particular weather station is likely to vastly underestimate the value of a future report from one of those weather stations that falls at the right place at the right time, and similarly the value of the intervention forecaster, who often has to do nothing, literally. The second is that the Burns' Day forecast illustrates something a bit different from the butterfly effect. Mathematical models allow us to worry about what the real future will bring *not* by considering possible worlds, of which there may be only one, but by contrasting different simulations of our model, of which there can be as many as we can afford. As Burns might appreciate, science gives us new ways to guess and new things to fear. The butterfly effect contrasts different worlds: one world with the nail and another world without that nail. The ***Burns effect*** places the focus firmly on us and our attempts to make rational decisions in the real world given only collections of different simulations under various imperfect models. The failure to distinguish between reality and our models, between observations and mathematics, arguably between an empirical fact and scientific fiction, is the root of much confusion regarding chaos both by the public and among scientists. It was research into nonlinearity and chaos that clarified yet again how import this distinction remains. In Chapter 10, we will return to take a deeper look at how today's weather forecasters would have used insights from their understanding of chaos when making a forecast for this event.

We have now touched on the three properties found in chaotic mathematical systems: chaotic systems are nonlinear, they are deterministic, and they are unstable in that they display sensitivity to initial condition. In the chapters that follow we will constrain them further, but our real interests lie not only in the mathematics of chaos, but also in what it can tell us about the real world.

Chaos and the real world: predictability and a 21st-century demon

There is no more greater an error in science, than to believe that just because some mathematical calculation has been completed, some aspect of Nature is certain.

Alfred North Whitehead (1953)

What implications does chaos hold for our everyday lives? Chaos impacts the ways and means of weather forecasting, which affect us directly through the weather, and indirectly through economic consequences both of the weather and of the forecasts themselves. Chaos also plays a role in questions of climate change and our ability to foresee the strength and impacts of global warming. While there are many other things that we forecast, weather and climate can be used to represent short-range forecasting and long-range modelling, respectively. 'When is the next solar eclipse?' would be a weather-like question in astronomy, while 'Is the solar system stable?' would be a climate-like question. In finance, when to buy 100 shares of a given stock is a weather-like question, while a climate-like question might address whether to invest in the stock market or real estate.

Chaos has also had a major impact on the sciences, forcing a close re-examination of what scientists mean by the words 'error' and 'uncertainty' and how these meanings change when applied to our

16

world and our models. As Whitehead noted, it is dangerous to interpret our mathematical models as if they somehow governed the real world. Arguably, the most interesting impacts of chaos are not really new, but the mathematical developments of the last 50 years have cast many old questions into a new light. For instance, what impact would uncertainty have on a 21st-century incarnation of Laplace's demon which could not escape observational noise?

Consider an intelligence that knew all the laws of nature precisely and had good, but imperfect, observations of an isolated chaotic system over an arbitrarily long time. Such an agent – even if sufficiently vast to subject all this data to computationally exact analysis – could not determine the current state of the system and thus the present, as well as the future, would remain uncertain in her eyes. While our agent could not predict the future exactly, the future would hold no real surprises for her, as she could see what could and what could not happen, and would know the probability of any future event: the predictability of the world she could see. Uncertainty of the present will translate into well-quantified uncertainty in the future, *if* her model is perfect.

In his 1927 Gifford Lectures, Sir Arthur Eddington went to the heart of the problem of chaos: some things are trivial to predict, especially if they have to do with mathematics itself, while other things seem predictable, sometimes:

> A total eclipse of the sun, visible in Cornwall is prophesied for 11 August 1999 ... I might venture to predict that $2 + 2$ will be equal to 4 even in 1999 ... The prediction of the weather this time next year ... is not likely to ever become practicable ... We should require extremely detailed knowledge of present conditions, since a small local deviation can exert an ever-expanding influence. We must examine the state of the sun ... be forewarned of volcanic eruptions, ... , a coal strike ... , a lighted match idly thrown away ...

Our best models of the solar system are chaotic, and our best models of the weather appear to be chaotic: yet why was Eddington confident in 1928 that the 1999 solar eclipse would occur? And equally confident that no weather forecast a year in advance would ever be accurate? In Chapter 10 we will see how modern weather forecasting techniques designed to better cope with chaos helped me to see that solar eclipse.

When paradigms collide: chaos and controversy

One of the things that has made working in chaos interesting over the last 20 years has been the friction generated when different ways of looking at the world converge on the same set of observations. Chaos has given rise to a certain amount of controversy. The studies that gave birth to chaos have revolutionized not only the way professional weather forecasters forecast but even what a forecast consists of. These new ideas often run counter to traditional statistical modelling methods, and still produce both heat and light on how best to model the real world. This battle is broken into skirmishes by the nature of the field and our level of understanding in the particular system of which a question is asked, be it the population of voles in Scandinavia, a mathematical calculation to quantify chaos, the number of spots on the Sun's surface, the price of oil delivered next month, tomorrow's maximum temperature, or the date of the last ever solar eclipse.

The skirmishes are interesting, but chaos offers deeper insights even when both sides are fighting for traditional advantage, say, the 'best' model. Here studies of chaos have redefined the high ground: today we are forced to reconsider new definitions for what constitutes the best model, or even a 'good' model. Arguably, we must give up the idea of approaching Truth, or at least define a wholly new way of measuring our distance from it. The study of chaos motivates us to establish utility without any hope of achieving perfection, and to give up many obvious home truths of forecasting, like the naïve idea that a good forecast consists of a prediction that is close to the target. This did not appear naïve before we understood the implications of chaos.

La Tour's realistic vision of science in the real world

To close this chapter, we illustrate how chaos can force us to reconsider what constitutes a good model, and revise our beliefs as to what is ultimately responsible for our forecast failures. This impact is felt by scientists and mathematicians alike, but the reconsideration will vary depending on the individual's point of view and the empirical system under study. The situation is nicely personified in Figure 5, a French baroque painting by Georges de la Tour showing a card game from the 17th century. La Tour was arguably a realist with a sense of humour. He was fond of fortune telling and games of chance, especially those in which chance played a somewhat lesser role than the participants happened to believe. In theory, chaos can play exactly this role. We will interpret

5. *The Cheat with the Ace of Diamonds*, **by Georges de la Tour, painted about 1645**

this painting to show a mathematician, a physicist, a statistician, and a philosopher engaged in an exercise of skill, dexterity, insight, and computational prowess; this is arguably a description for doing science, but the task at hand here is a game of poker. Exactly who is who in the painting will remain open, as we will return to these personifications of natural science throughout the book. The insights chaos yields vary with the perspective of the viewer, but a few observations are in order.

The impeccably groomed young man on the right is engaged in careful calculations, no doubt a probability forecast of some nature; he is currently in possession of a handsome collection of gold coins on the table. The dealer plays a critical role, without her there is no game to be played; she provides the very language within which we communicate, yet she seems to be in nonverbal communication with the handmaiden. The role of the handmaiden is less clear; she is perhaps tangential, but then again the provision of wine will influence the game, and she herself may feature as a distraction. The roguish character in ramshackle dress with bows untied is clearly concerned with the real world, not mere appearances in some model of it; his left hand is extracting one of several aces of diamonds from his belt, which he is about to introduce into the game. What then do the 'probabilities' calculated by the young man count for, if, in fact, he is not playing the game his mathematical model describes? And how deep is the insight of our rogue? His glance is directed to us, suggesting that he knows we can see his actions, perhaps even that he realizes that he is in a painting?

The story of chaos is important because it enables us to see the world from the perspective of each of these players. Are we merely developing the mathematical language with which the game is played? Are we risking economic ruin by over-interpreting some potentially useful model while losing sight of the fact that it, like all models, is imperfect? Are we only observing the big picture, not entering the game directly but sometimes providing an interesting distraction? Or are we manipulating those things we can change, acknowledging the risks of model inadequacy, and perhaps even our own limitations, due to being within the system? To answer these questions we must first examine several of the many jargons of science in order to be able to see how chaos emerged from the noise of traditional linear statistics to vie for roles both in understanding and in predicting complicated real-world systems. Before the nonlinear dynamics of chaos were widely recognized within science, these questions fell primarily in the domain of the philosophers; today they reach out via our mathematical models to physical scientists and working forecasters, changing the statistics of decision support and even impacting politicians and policy makers.

Chapter 2
Exponential growth, nonlinearity, common sense

One of the most pervasive myths about chaotic systems is that they are impossible to predict. To expose the fallacy of this myth, we must understand how uncertainty in a forecast grows as we predict further and further into the future. In this chapter we investigate the origin and meaning of ***exponential growth***, since on average a small uncertainty will grow exponentially fast in a chaotic system. There is a sense in which this phenomenon really does imply a 'faster' growth of uncertainty than that found in our traditional ideas of how error and uncertainty grow as we forecast further into the future. Nevertheless, chaos can be easy to predict, sometimes.

Chess, rice, and Leonardo's rabbits: exponential growth

An oft-told story about the origin of the game of chess illustrates nicely the speed of exponential growth. The story goes that a king of ancient Persia was so pleased when first presented with the game that he wanted to reward the game's creator, Sissa Ben Dahir. A chess board has 64 squares arranged in an 8 by 8 pattern; for his reward, Ben Dahir requested what seemed a quite modest sum of rice determined using the new chess board: one grain of rice was to be put on the first square of the board, two to be put on the second, four for the third, eight for the fourth, and so on, doubling the number on each square until the 64th was reached. A mathematician will often call any rule for generating one number from another one a mathematical **map**, so we'll refer to this simple rule ('double the current value to generate the next value') as the *Rice Map*.

Before working out just how much rice Ben Dahir has asked for, let us consider the case of linear growth where we have one grain on the first square, two on the second square, three on the third, and so on until we need 64 for the last square. In this case we have a total of $64 + 63 + 62 + \ldots + 3 + 2 + 1$, or around 1,000 grains. Just for comparison, a 1 kilogram bag of rice contains a few tens of thousands of grains.

The Rice Map requires one grain for the first square, then two for the second, four for the third, then 8, 16, 32, 64, and 128 for the last square of the first row. On the third square of the second row, we pass 1,000 and before the end of the second row there is a square which exhausts our bag of rice. To fill the next square alone will require another entire bag, the following square two bags, and so on. Some square in the third row will require a volume of rice comparable to a small house, and we will have enough rice to fill the Royal Albert Hall well before the end of the fifth row. Finally, the 64th square alone will require billions and billions, or to be exact, 2^{63} ($= 9, 223, 372, 036, 854, 775, 808$) grains, for a total of 18,446,744,073,709,551,615 grains. That is a non-trivial quantity of rice! It is something like the entire world's rice production over two millennia. Exponential growth quickly grows out of all proportion.

By comparing the amount of rice on a given square in the case of linear growth with the amount of rice on the same square in the case of exponential growth, we quickly see that exponential is much faster than linear growth: on the fourth square we already have twice as many grains in the exponential case as in the linear case (8 in the first, only 4 in the second), and by the eighth square, at the end of the first row, the exponential case has 16 times more! Soon thereafter we have the astronomical numbers.

Of course, we hid the values of some **parameters** in the example above: we could have made the linear growth faster by adding not one additional grain for each square, but instead, say, 1,000 additional grains. This parameter, the number of additional grains, defines the constant of proportionality between the number of the square and the number of grains on that square, and gives us the slope of the linear relationship between them. There is also a parameter in the exponential case: on each step we increased the number of grains by a factor of two, but it could have been a factor of three, or a factor of one and a half.

One of the surprising things about exponential growth is that *whatever* the values of these parameters, there will come a time at which exponential growth surpasses *any* linear growth, and will soon thereafter dwarf linear growth, no matter how fast the linear growth is. Our ultimate interest is not in rice on a chess board, but in the dynamics of uncertainty in time. Not just the growth of a population, but the growth of our uncertainty in a forecast of the future size of that population. In the forecasting context, there will come a time at which an exponentially growing uncertainty which is very small today will surpass a linearly growing uncertainty which is today much larger. And the same thing happens when contrasting exponential growth with growth proportional to the square of time, or to the cube of time, or to time raised to any power (in symbols: steady exponential growth will eventually surpass the growth proportional to t^2 or t^3 or t^n for any value of n.). It is for this reason among others that exponential growth is mathematically distinguished, and taken to provide a benchmark for defining chaos. It has also contributed to the widespread but fundamentally mistaken impression that chaotic systems are hopelessly unpredictable. Ben Dahir's chess board illustrates that there is a deep sense in which exponential growth is faster than linear growth. To place this in the context of forecasting, we move forward a few hundred years in time and a few hundred miles northwest, from Persia to Italy.

At the beginning of the 13th century, Leonardo of Pisa posed a question of population dynamics: given a newborn pair of rabbits in a large, lush, walled garden, how many pairs of rabbits will we have in one year if their nature is for each mature pair to breed and produce a new pair every month, and newborn rabbits mature in their second month? In the first month we have one juvenile pair. In the second month this pair matures and breeds to produce a new pair in the third month. So in the third month, we have one mature pair and one newborn pair. In the fourth month we once again have one new born pair from the original pair of rabbits and now two mature pairs for a total of three pairs. In the fifth month, two new pairs are born (one from each mature pair), and we have three mature pairs for a total of five pairs. And so on.

So what does this 'population dynamic' look like? In the first month we have one immature pair, in the second month we have one mature pair, in the third month we have one mature pair and a new immature pair, in the fourth month we have two mature pairs and one immature pair, in the fifth month we have three mature pairs and two immature.

If we count up all the pairs each month, the numbers are 1, 1, 2, 3, 5, 8, 13, 21 . . . . Leonardo noted that the next number in the series is always the sum of the previous two numbers $(1 + 1 = 2, 2 + 1 = 3, 3 + 2 = 5, \dots)$ which makes sense, as the previous number is the number we had last month (in our model all rabbits survive no matter how many there are), and the penultimate number is the number of mature pairs (and thus the number of new pairs arriving this month).

Now it gets a bit tedious to write 'and in the sixth month we have 12 pairs of rabbits', so scientists often use a short-hand X for the number of pairs of rabbits and X_6 to denote the number of pairs in month six. And since the series 1, 1, 2, 3, 5, 8, . . . reflects how the population of rabbits evolves in time, this series and others like it are called *time series*. The Rabbit Map is defined by the rule:

Add the previous value of X to the current value of X, and take the sum as the new value of X.

The numbers in the series 1, 1, 2, 3, 5, 8, 13, 21, 34 . . . are called Fibonacci numbers (Fibonacci was a nickname of Leonardo of Pisa), and they arise again and again in nature: in the structure of sunflowers, pine cones, and pineapples. They are of interest here because they illustrate exponential growth in time, almost. The crosses in Figure 6 are Fibonacci's points – the rabbit population as a function of time – while the solid line reflects two raised to the power λt, or in symbols $2^{\lambda t}$, where t is the time in months and λ is our first exponent. Exponents which multiply time in the superscript are a useful way of quantifying uniform exponential growth. In this case, λ is equal to the logarithm of a number called the golden mean, a very special number which is discussed in the *Very Short Introduction to Mathematics*.

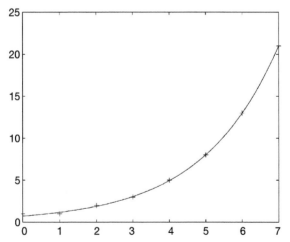

6. The series of crosses showing the number of pairs of rabbits each month (Fibonacci numbers); the smooth curve they lie near is the related exponential growth

The first thing to notice about Figure 6 is that the points lie close to the curve. The exponential curve is special in mathematics because it reflects a function whose increase is proportional to its current value. The larger it gets, the faster it grows. It makes sense that something like this function would describe the dynamics of Leonardo's rabbit population since the number of rabbits next month is more or less proportional to the number of rabbits this month. The second thing to notice about the figure is that the points do *not* lie on the curve. The curve is a good **model** for Fibonacci's Rabbit Map, but it is not perfect: at the end of each month the number of rabbits is always a whole number and, while the curve may be close to the correct whole number, it is not exactly equal to it. As the months go by and the population grows, the curve gets closer and closer to each Fibonacci number, but it never reaches them. This concept of getting closer and closer but never quite arriving is one that will come up again and again in this book.

So how can Leonardo's rabbits help us to get a feel for the growth of forecast uncertainty? Like all observations, counting the number of rabbits in a garden is subject to error; as we saw in Chapter 1, observational uncertainties are said to be caused by noise. Imagine that Leonardo failed to notice a pair of mature rabbits also in the garden in the first month; in that case, the number of pairs actually in the garden would have been 2, 3, 5, 8, 13, . . . The error in the original forecast (1, 1, 2, 3, 5, 8 . . .) would be the difference between the Truth and that forecast, namely: 1, 2, 3, 5 . . . (again, the Fibonacci series). In month 12, this error has reached a very noticeable 146 pairs of rabbits! A small error in the initial number of rabbits results in a very large error in the forecast. In fact, the error is growing exponentially in time. This has many implications.

Consider the impact of the exponential error growth on the uncertainty of our forecasts. Let us again contrast linear growth and exponential growth. Let's assume that, for a price, we can reduce the uncertainty in the initial observation that we use in generating our forecast. If the error growth is linear, and we reduce our initial uncertainty by a factor of ten, then we can forecast the system ten times longer before our uncertainty exceeds the same threshold. If we reduce the initial uncertainty by a factor of 1,000, then we can get forecasts of the same quality 1,000 times longer. This is an advantage of linear models. Or, more accurately, this is an apparent advantage of studying only linear systems. By contrast, if the model is nonlinear and the uncertainty grows exponentially, then we may reduce our initial uncertainty by a factor of ten yet only be able to forecast twice as long with the same accuracy. In that case, *assuming* the exponential growth in uncertainty is uniform in time, reducing the uncertainty by a factor of 1,000 will only increase our forecast range at the same accuracy by a factor of eight. Now reducing the uncertainty in a measurement is rarely free (we have to hire someone else to count the rabbits a second time), and large reductions of uncertainty can be expensive, so when uncertainty grows exponentially fast, the cost sky-rockets. Attempting to achieve our forecast goals by reducing uncertainty in initial conditions can be tremendously expensive.

Luckily, there is an alternative that allows us to accept the simple fact that we can never be certain that any observation has not been corrupted by noise. In the case of rabbits or grains of rice, it seems there really is a fact of the matter, a whole number that reflects the correct answer. If we reduce the uncertainty in this initial condition to zero then we can predict without error. But can we ever really be certain of the initial condition? Might there not be another bunny hiding in the noise? While our best guess is that there is one pair in the garden, there might be two, or three, or more (or perhaps zero). When we are uncertain of the initial condition, we can examine the diversity of forecasts under our model by making an ensemble of forecasts: one forecast started from each initial condition we think plausible. So one member of the ensemble will start with X equal to one, another ensemble member will start with X equals two, and so on. How should we divide our limited resources between computing more ensemble members and making better observations of the current number of rabbits in the garden?

In the Rabbit Map, differences between the forecasts of different members of the ensemble will grow exponentially fast, but with an ensemble forecast we can see just how different they are and use this as a measure of our uncertainty in the number of rabbits we expect at any given time. In addition, if we carefully count the number of rabbits after a few months, we can all but rule out some of the individual ensemble members. Each of these ensemble members was started from some estimate of the number of rabbits that were in the garden originally, so ruling an ensemble member out in effect gives us more information about the original number of rabbits. Of course, this information need only prove accurate if our model is literally perfect, meaning, in this case, that our Rabbit Map captures the reproductive behaviour and longevity of our rabbits exactly. But if our model is perfect, then we can use future observations to learn about the past; this process is called *noise reduction*. If it turns out that our model is not perfect, then we may end up with incoherent results.

But what if we were measuring something that is not a whole number, like temperature, or the position of a planet? And is temperature in an imperfect weather model exactly the same thing as temperature in the real world? It was these questions that initially interested our philosopher in chaos. First, we should consider the more pressing question of why rabbits have not taken over the world in the 9,000 months since 1202?

Stretching, folding, and the growth of uncertainty

The study of chaos lends credence to the meteorological maxim that no forecast is complete without a useful estimate of forecast uncertainty: if we know our initial condition is uncertain then we are not only interested in the prediction *per se*, but equally in learning what the likely forecast error will be. Forecast error for any

Exponential growth: an example from Miss Nagel's third grade class

A few months ago, I received an email written by an old friend of mine from elementary school. It contained another email that had originated from a third grader in North Carolina whose class was studying geography. It requested that everyone who read the email send a reply to the school stating where they lived, and the class would locate that place on a school globe. It also requested that each reader pass on the email to ten friends.

I did not forward the message to anyone, but I did write an email to Miss Nagel's class stating that I was in Oxford, England. I also suggested that they tell their mathematics teacher about their experiment and use it as an example to illustrate exponential growth: if they sent the message to ten people, and the next day each of them sent it to ten more people, that would be 100 on day three, 1,000 on day four, and more emails than there are email addresses within a week or so. In a real system, exponential growth cannot go on forever: eventually we run out of rice, or garden space, or new email addresses. It is often the resources that limit growth: even a lush garden provides only a finite amount of rabbit food. There are limits to growth which bound populations, if not our models of populations.

I never found out whether Miss Nagel's class learned their lesson in exponential growth. The only answer I ever received was an automated reply stating that the school's email in-box had exceeded its quota and had been closed.

real system should not grow without limit; even if we start with a small error like one grain or one rabbit, the forecast error will not grow arbitrarily large (unless we have a very naïve forecaster), but will saturate near some limiting value, as would the population itself. Our mathematician has a way to avoid ludicrously large forecast errors (other than naïveté), namely by making the initial uncertainty ***infinitesimally*** small – smaller than any number you can think of, yet greater than zero. Such an uncertainty will stay infinitesimally small for all time, even if it grows exponentially fast.

Physical factors, like the total amount of rabbit food in the garden or the amount of disk space on an email system, limit growth in practice. The limits are intuitive even if we do not know exactly what causes them: I think I have lost my keys in the car park; of course they might be several miles from there, but it is exceedingly unlikely that they are farther away than the moon. I do not need to understand or believe the laws of gravity to appreciate this. Similarly, weather forecasters are rarely more than 100 degrees off, even for a forecast one year in advance! Even inadequate models can usually be constrained so that their forecast errors are bounded.

Whenever our model goes into never-never land (suggesting values where no data have ever gone before), then something is likely to give, unless something in our model has already broken. Often, as our uncertainty grows too large, it starts to fold back on itself. Imagine kneading dough, or a toffee machine continuously stretching and folding toffee. An imaginary line of toffee connecting two very nearby grains of sugar will grow longer and longer as these two grains separate under the action of the machine, but before it becomes bigger than the machine itself, this line will be folded back into itself, forming a horrible tangle. The distance between the grains of sugar will stop growing, even as the string of toffee connecting them continues to grow longer and longer, becoming a more and more complicated tangle. The toffee machine gives us a way to envision limits to the growth of prediction error whenever our model is perfect. In this case, the error is the growing *distance* between the True state and our best guess of that state: any exponential growth of error would correspond only to the rapid initial growth of the string of toffee. But if our forecasts are not going to zoom away towards infinity (the toffee must stay in the machine, only a finite number of rabbits will fit in the garden, and the like), then eventually the line connecting Truth and our forecast will be folded over on itself. There is simply nowhere else for it to grow into. In many ways, identifying the movement of a grain of sugar in the toffee machine with the evolution of the state of a chaotic system in three dimensions is a useful way to visualize chaotic motion.

We want to require a sense of containment for chaos, since it is hardly surprising that it is difficult to predict things that are flying apart to infinity, but we do not want to impose so strict a condition as requiring a forecast to never exceed some limited value, no matter how big that value might be. As a compromise, we require the system to come back to the vicinity of its current state at some point in the future, and to do so again and again. It can take as long as it wants to come back, and we can define coming back to mean returning closer to the current point than we have ever seen it return before. If this happens, then the trajectory is said to be *recurrent*. The toffee again provides an analogy: if the motion was chaotic and we wait long enough, our two grains of sugar will again come back close together, and each will pass close to where it was at the beginning of the experiment, assuming no one turns off the machine in the meantime.

Chapter 3

Chaos in context: determinism, randomness, and noise

All linear systems resemble one another, each nonlinear system is nonlinear in its own way.

After Tolstoy's *Anna Karenina*

Dynamical systems

Chaos is a property of dynamical systems. And a dynamical system is nothing more than a source of changing observations: Fibonacci's imaginary garden with its rabbits, the Earth's atmosphere as reflected by a thermometer at London's Heathrow airport, the economy as observed through the price of IBM stock, a computer program simulating the orbit of the moon and printing out the date and location of each future solar eclipse.

There are at least three different kinds of dynamical systems. Chaos is most easily defined in *mathematical dynamical systems*. These systems consist of a rule: you put a number in and you get a new number out, which you put back in, to get yet a newer number out, which you put back in. And so on. This process is called ***iteration***. The number of rabbits each month in Fibonacci's imaginary garden is a perfect example of a time series from this kind of system. A second type of dynamical system is found in the empirical world of the physicist, the biologist, or the stock market trader. Here, our sequence of observations consists of noisy measurements of reality, which are fundamentally different from the noise-free numbers of the Rabbit Map. In these *physical dynamical systems* – the Earth's atmosphere and Scandinavia's vole population, for example – numbers represent the state, whereas in the Rabbit Map they *were* the state. To avoid needless confusion, it is useful to distinguish a third case when a digital computer performs the arithmetic specified by a mathematical dynamical system; we will call this a *computer simulation* – computer programs that produce TV weather forecasts are a common example. It is important to remember that these are different *kinds* of systems and that each is a different beast: our best equations for the weather differ from our best computer models based on those equations, and both of these systems differ from the real thing the Earth's atmosphere itself. Confusingly, the numbers from each of our three types of systems are called time series, and we must constantly struggle to keep in mind the distinction between what these are time series of: a number of imaginary rabbits, the True temperature at the airport (if such a thing exists), a measurement representing that temperature, and a computer simulation of that temperature.

The extent to which these differences are important depends on what we aim to do. Like la Tour's card players, scientists, mathematicians, statisticians, and philosophers each have different talents and aims. The physicist may aim to describe the observations with a mathematical model, perhaps testing the model by using it to predict future observations. Our physicist is willing to sacrifice mathematical tractability for physical relevance. Mathematicians like to prove things that are true for a wide range of systems, but they value proof so highly that they often do not care how widely they must restrict that range to have it; one should almost always be wary whenever a mathematician is heard to say '***almost every***'. Our physicist must be careful not to forget this and confuse mathematical utility with physical relevance; physical intuitions should not be biased by the properties of 'well-understood' systems designed only for their mathematical tractability.

Our statistician is interested in describing interesting statistics from the time series of real observations and in studying the properties of dynamical systems that generate time series which look like the observations, always taking care to make as few assumptions as possible. Finally, our philosopher questions the relationships among the underlying physical system that we claim generated the observations, the observations themselves, and the mathematical models or statistical techniques that we created to analyse them. For example, she is interested in what we can know about the relationship between the temperature we measure and the true temperature (if such a thing exists), and in whether the limits on our knowledge are merely practical difficulties we might resolve or limits in principle that we can never overcome.

Mathematical dynamical systems and attractors

We commonly find four different types of behaviour in time series. They can (i) grind to a halt and more or less repeat the same fixed number over and over again, (ii) bounce around in a closed loop like a broken record, periodically repeating the same pattern: exactly the same series of numbers over and over, (iii) move in a loop that has more than one period and so does not quite repeat exactly but comes close, like the moment of high tide drifting through the time of day, or (iv) forever jump about wildly, or perhaps even calmly, displaying no obvious pattern. The fourth type looks random, yet looks can be deceiving. Chaos can look random but it is not random. In fact, as we have learned to see better, chaos often does not even look all that random to us anymore. In the next few pages we will introduce several more maps, though perhaps without the rice or rabbits. We need these maps in order to generate interesting artefacts for our tour in search of the various types of behaviour just noted. Some of these maps were generated by mathematicians for this very purpose, although our physicist might argue, with reason, that a given map was derived by simplifying physical laws. In truth, the maps are simple enough to have each come about in several different ways.

Before we can produce a time series by iterating a map, we need some number to start with. This first number is called an *initial condition*, an initial **state** that we define, discover, or arrange for our system to be. As in Chapter 2, we adopt the symbol X as shorthand for a state of our system. The collection of all possible states X is called the **state space**. For Fibonacci's imaginary rabbits, this would be the set of all whole numbers. Suppose our time series is from a model of the average number of insects per square mile at mid-summer each year. In that case, X is just a number and the state space, being the collection of all possible states, is then a line. It sometimes takes more than one number to define the state, and if so X will have more than one component. In predator-prey models, for instance, the populations of both are required and X has two components: it is a vector. When X is a vector containing both the number of voles (prey) and the number of weasels (predators) on the first of January each year, then the state space will be a two-dimensional surface – a plane – that contains all pairs of numbers. If X has three components (say, voles, weasels, and annual snowfall), then the state space is a three-dimensional space containing all triplets of numbers. Of course, there is no reason to stop at three components; although the pictures become more challenging to draw in higher dimensions, modern weather models have over 10,000,000 components. For a mathematical system, X can even be a continuous field, like the height of the surface of the ocean or the temperature at every point on the surface of the Earth. However, our observations of physical systems will never be more complicated than a vector, and since we will only measure a finite number of things, our observations will always be finite-dimensional vectors. For the time being, we will consider the case in which X is a simple number, such as one-half.

Recalling that a mathematical map is just a rule that transforms one set of values into the next set of values, you can define the **Quadrupling Map** by the rule:

Multiply X by four to form the new value of X.

Given an initial condition, like X equals one-half, this mathematical dynamical system produces a time series of values of X, in this case ½ × 4 = 2, 2 × 4 = 8, 8 × 4 = 32 . . . and the time series is 0.5, 2, 8, 32, 128, 512, 2048 . . . And so on. This series just gets bigger and bigger and, dynamically speaking, that is not so interesting. If a time series of X grows without limit like this one does, we call it *unbounded*. In order to get a dynamical system where X is bounded, we'll take a second example, the **Quartering Map**:

Take X divided by four as the new X

Starting at X = ½ yields the time series 1/8, 1/32, 1/128, ... At first sight, this is not very exciting since X rapidly shrinks towards zero. But in fact, the Quartering Map has been carefully designed to illustrate special mathematical properties. The origin – the state X = 0 – is a *fixed point*: if we start there we will never leave, since zero divided by four is again zero. The origin is also our first *attractor*; under the Quartering Map the origin is the inevitable if unreachable destination: if we start with some other value of X, we never actually make it to the attractor, although we get close as the number of iterations increases without limit. How close? Arbitrarily close. As close as you like. *Infinitesimally* close, meaning closer than any number you can name. Name a number, any number, and we can work out how many iterations are required after which X will remain closer to zero than that number. Getting arbitrarily close to an attractor as time goes on while never quite reaching it is a common feature of many time series from nonlinear systems. The pendulum provides a physical analogue: each swing will be smaller than the last, an effect we blame on air resistance and friction. The analogue of the attractor in this case is the motionless pendulum hanging straight down. We will have more to say about attractors after we have added a few more dynamical systems to our menagerie.

In the **Full Logistic Map**, time series from almost every X bounces around irregularly between zero and one forever:

Subtract X^2 from X, multiply the difference by four and take the result as the new X.

If we multiply components of state variables by other components, things become nonlinear. What is the time series in this case if we again start with X equals one-half? Starting with ½, X minus X^2 is ¼, times four is one, so our new value is one. Continuing with X now equal to one, we have X minus X^2 is zero. But four times zero is always zero, so we'll get zeros forever. And our time series is 0.5, 1, 0, 0, 0 . . . This does not blow up, but it is hardly exciting; recall the warning about 'almost every'.

The order of the numbers in a time series is important, whether the series reflects monthly values of Fibanocci's rabbits or iterations of the Full Logistic Map. Using the short-hand suggested in Chapter 2, we will write X_5 for the fifth new value of X, and X_0 for the initial state (or observation), and in general X_i for the ith value. Whether we are iterating the map or taking observations, i is always an integer and is often called 'time'.

In the Full Logistic Map with X_0 is equal to 0.5, X_1 is equal to 1, X_2 is 0, X_3 is 0, X_4 is 0, and X_i will be zero for all i greater than four as well. So the origin is again a fixed point. But under the Full Logistic Map small values of X grow (you can check this with a hand calculator), X = 0 is unstable and so the origin is not an attractor. A time series started near the origin is in fact unlikely to take one of the first three options noted at the opening of this section, but to bounce about chaotically forever.

Figure 7 shows a time series starting near X_0 equals 0.876; it represents a chaotic time series from the Full Logistic Map. But look at it closely: does it really look completely unpredictable? It looks like small values of X are followed by small values of X, and that there is a tendency for the time series to linger whenever it is near three-quarters. Our physicist would look at this series and expect it to be predictable at least sometimes, while, after a few

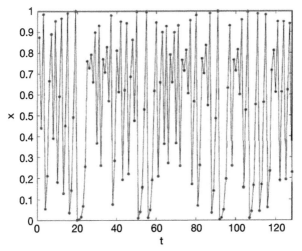

7. A chaotic time series from the Full Logistic Map starting near X_0 equals 0.876. Note the series is visibly predictable whenever X is near zero and three-quarters

calculations, our statistician might even declare it random. Although we can see this structure, the most common statistical tests cannot.

A menagerie of maps

The rule that defines a map can be stated either in words, or as an equation, or in a graph. Each panel of Figure 8 defines the rule graphically. To use the graph, find the current value of X on the horizontal axis, and then move directly upward until you hit the curve; the value of this point on the curve on the vertical axis is the new value of X. The Full Logistic Map is shown graphically in Figure 8 (b), while the Quarter Map is in panel (a).

An easy way of using the graph to see if a fixed point is unstable is to look at the slope of the map at the fixed point: if the slope is steeper than 45 degrees (either up or down); then the fixed point is

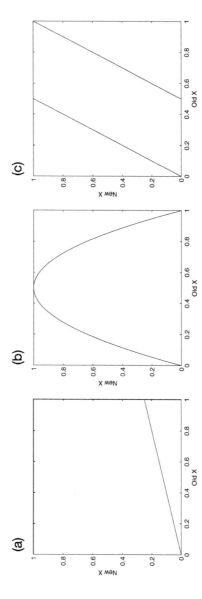

8. Graphical presentation of the (a) Quarter Map, (b) Full Logistic Map, (c) Shift Map, (d) Tent Map, (e) Tripling Tent Map, and (f) the Moran-Ricker Map

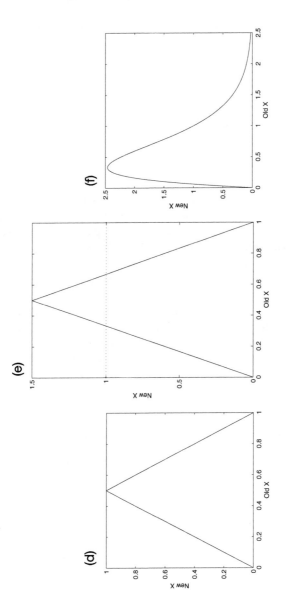

unstable. In the Quartering Map the slope is less than one everywhere, while for the Full Logistic Map the slope near the origin is greater than one. Here small but non-zero values of X grow with each iteration but only as long as they stay sufficiently small (the slope near ½ is zero). As we will see below, for *almost every* initial condition between zero and one, the time series displays true mathematical **chaos**. The Full Logistic Map is pretty simple; chaos is pretty common.

To see if a mathematical system is **deterministic** merely requires checking carefully whether carrying out the rule requires a random number. If not, then the dynamical system is deterministic: every time we put the same value of X in, we get the same new value of X out. If the rule requires (really requires) a random number, then the system is random, also called *stochastic*. With a stochastic system, even if we iterate *exactly* the same initial condition we expect the details of the next value of X and thus the time series to be different. Looking back at their definitions, we see that the three maps defined above are each deterministic; their future time series is completely determined by the initial condition, hence the name 'deterministic system'. Our philosopher would point out that just knowing X is not enough, we also need to know the mathematical system and we have to have the power to do exact calculations with it. These were the three gifts Laplace ensured his demon possessed 200 years ago.

Our first stochastic dynamical system is the **AC Map**:

> Divide X by four, then subtract ½ and add a random number R to get the new X.

The AC Map is a stochastic system since applying the rule requires access to a supply of random numbers. In fact, the rule above is incomplete, since it does not specify how to get R. To complete the definition we must add something like: for R on each iteration, pick a number between zero and one in a manner that each number is equally likely to be chosen, which implies that R will be uniformly distributed between zero and one and that the probability of the next value of R falling in an interval of values is proportional to the width of that interval.

What rule do we use to pick R? It could not be a deterministic rule, since then R would not be random. Arguably, there is no finite rule for generating values of R. This has nothing to do with needing uniform numbers between zero and one. We'd have the same problem if we wanted to generate random numbers which mimicked Galton's 'bell-shape' distribution. We will have to rely on our statistician to somehow get us the random numbers we need; hereafter we'll just state whether they have a uniform distribution or the bell-shaped distribution.

In the AC Map, each value of R is used within the map, but there is another class of random maps – called Iterated Function Systems, or IFS for short – which appear to use the value of R not in a formula but to make a decision as to what to do. One example is the Middle Thirds IFS Map, which will come in handy later when we try to work out the properties of maps from the time series that they generate. The Middle Thirds IFS Map is:

> Take a random number R from a uniform distribution between zero and one.
> If R is less than a half, take X/3 as the new X
> Otherwise take 1 – X/3 as the new X.

So now we have a few mathematical systems, and we can easily tell if they are deterministic or stochastic. What about computer simulations? Digital computer simulations are always deterministic. And as we shall see in Chapter 7, the time series from a digital computer is either on an endless loop of values repeating itself periodically, over and over again, or it is on its way towards such a loop. This first part of a time series in which no value is repeated, the trajectory is evolving towards a ***periodic loop*** but has not reached it, is called a ***transient***. In mathematical circles, this word is something of an insult, since mathematicians prefer to work with long-lived things, not mere transients. While mathematicians avoid transients, physical scientists may never see anything else and, as it turns out, digital computers cannot maintain them. The digital computers that have proven critical in advancing our understanding of chaos cannot, ironically, display true mathematical chaos themselves. Neither can a digital computer generate random numbers. The so-called random number generators on digital computers and hand calculators are, in fact, only pseudo-random number generators; one of the earliest of these generators was even based on the Full Logistic Map! The difference between mathematical chaos and computer simulations, like that between random numbers and pseudo-random numbers, exemplifies the difference between our mathematical systems and our computer simulations.

The maps in Figure 8 are not there by chance. Mathematicians often construct systems in such a way that it will be relatively simple for them to illustrate some mathematical point or allow the application of some specific manipulation – a word they sometimes use to obscure technical sleight of hand. The really complicated maps – including the ones used to guide spacecraft and the ones called 'climate models', and the even bigger ones used in numerical weather prediction – are clearly constructed by physicists, not mathematicians. But they all work the same way: a value of X goes in and a new value X comes out. The mechanism is exactly the same as in the simple maps defined above, even if X might have over 10,000,000 components.

Parameters and model structure

The rules that define the maps above each involve numbers other than the state, numbers like four and one-half. These numbers are called **parameters**. While X changes with time, parameters remain fixed. It is sometimes useful to contrast the properties of time series generated using different parameter values. So instead of defining the map with a particular parameter value, like 4, maps are usually defined using a symbol for the parameter, say α. We can then contrast the behaviour of the map at α equals 4 with that at α = 2, or α = 3.569945, for example. Greek symbols are often used to clearly distinguish parameters from state variables. Rewriting the Full Logistic Map with a parameter yields one of the most famous systems of nonlinear dynamics: the **Logistic Map**:

Subtract X^2 from X, then multiply by α and take the result as the new X.

In physical models, parameters are used to represent things like the temperature at which water boils, or the mass of the Earth, or the speed of light, or even the speed with which ice 'falls' in the upper atmosphere. Statisticians often dismiss the distinction between the parameter and the state, while physicists tend to give parameters special status. Applied mathematicians, as it turns out, often force parameters towards the infinitely large or the infinitesimally small; it is easier, for example, to study the flow of air over an infinitely long wing. Once again, these different points of view each make sense in context. Do we require an exact solution to an approximate question, or an approximate answer to a particular question? In nonlinear systems, these can be very different things.

Attractors

Recall the Quartering Map, noting that after one iteration every point between zero and one will be between zero and one-quarter. Since all the points between zero and one-quarter are also between zero and one, none of these points can ever escape to values greater than one or less than zero. Dynamical systems in which, on average, line segments (or in higher dimensions, areas or volumes) shrink are called *dissipative*. Whenever a dissipative map translates a volume of state space completely inside itself, we know immediately that an attractor exists without knowing what it looks like.

Whenever α is less than four we can prove that the Logistic Map has an attractor by looking at what happens to all the points between zero and one. The largest new value of X we can get is the iteration of X equals one-half. (Can you see this in Figure 8?) This largest value is α/4, and as long as α is less than four this largest value is less than one. That means every point between zero and one iterates to a point between zero and α/4 and is confined there forever. So the system must have an attractor. For small values of α the point X equals zero is the attractor, just like in the Quartering Map. But if α is greater than one, then any value of X near zero will move away and the attractor is elsewhere. This is an example of a non-constructive proof: we can prove that an attractor exists but, frustratingly, the proof does not tell us how to find it nor give any hint of its properties!

Multiple time series of the Logistic Map for each of four different values of α are shown in Figure 9. In each panel, we start with 512 points taken at random between zero and one. At each step we move the entire ensemble of points forward in time. In the first step we see that all remain greater than zero, yet move away from X equals one never to return: we have an attractor. In (a) we see them all collapsing onto the period one loop; in (b) onto one of the two points in the period two loop; in (c) onto one of the four points of the period four loop. In (d), we can see that they are collapsing, but it is not clear what the period is. To make the dynamics more plainly visible, one member of our ensemble is chosen at random in the middle of the graph, and the points on its trajectory are joined by a line from that point forward. The period one loop (a) appears as a straight line, while (b) and (c) show the trajectories alternating between two or four points, respectively. While (d) first looks like a period four loop as well, but a closer look shows that there are many more than four options, and that while there is regularity in the order in which the bands of points are visited, no simple periodicity is visible.

To get a different picture of the same phenomena, we can examine many different initial conditions and different values for α at the same time, as shown in Figure 13 (page 63). In this three-dimensional view, the initial states can be seen randomly scattered on the back left of the box. At each iteration, they move out towards you and the points collapse towards the pattern shown in the previous two figures. The iterated initial random states are shown after 0, 2, 8, 32, 128, and 512 iterations; it takes some time for the transients to die away, but the familiar patterns can be seen emerging as the states reach the front of the box.

Tuning model parameters and structural stability

We can see now that a dynamical system has three components: the mathematical rule that defines how to get the next value, the parameter values, and the current state. We can, of course, change any of these things and see what happens, but it is useful to distinguish what type of change we are making. Similarly, we may have insight into the uncertainty in one of these components, and it is in our interest to avoid accounting for uncertainty in one component by falsely attributing it to another.

Our physicist may be looking for the 'True' model, or only just a useful one. In practice there is an art of 'tuning' parameter values. And while nonlinearity requires us to reconsider how we find 'good parameter values', chaos will force us to re-evaluate what we mean by 'good'. A very small difference in the value of a parameter which has an unnoticeable impact on the quality of a short-term forecast can alter the shape of an attractor beyond recognition. Systems in which this happens are called *structurally unstable*. Weather forecasters need not worry about this, but climate modellers must; as Lorenz noted in the 1960s.

A great deal of confusion has arisen from the failure to distinguish between uncertainty in the current state, uncertainty in the value of a parameter, and uncertainty regarding the model structure itself. Technically, chaos is a property of a

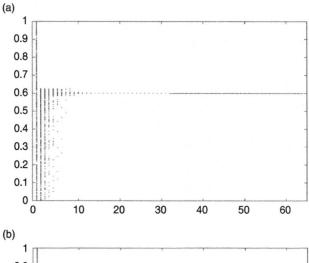

(a)

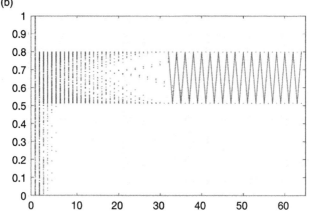

(b)

9. Each frame shows the evolution of 512 points, initially spread at random between zero and one, as they move forward under the Logistic Map. Each panel shows one of four different values of α, showing the collapse towards (a) a fixed point, (b) a period two loop, (c) a period four loop, and (d) chaos. The solid line starting at time 32 shows the trajectory of one point, in order to make the path on each attractor visible

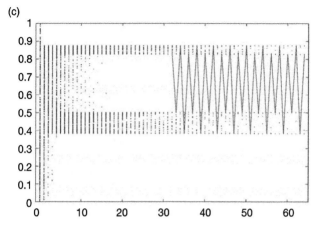

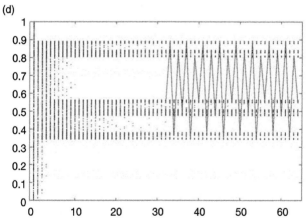

dynamical system with fixed equations (structure) and specified parameter values, so the uncertainty that chaos acts on is only the uncertainty in the initial state. In practice, these distinctions become blurred and the situation is much more interesting, and confused.

Statistical models of Sun spots

Chaos is only found in deterministic systems. But to understand its impact on science we need to view it against the background of traditional stochastic models developed over the past century. Whenever we see something repetitive in nature, periodic motion is one of the first hypotheses to be deployed. It can make you famous: Halley's comet, and the Wolf Sun spot number. In the end, the name often sticks even when we realize that the phenomenon is not really periodic. Wolf guessed that the Sun went through a cycle of about 11 years at a time when he had less than 20 years' data. Periodicity remains a useful concept even though it is impossible to prove a physical system is periodic regardless of how much data we take. So are the concepts of determinism and chaos.

The solar record showed correlations with weather, with economic activity, with human behaviour; even 100 years ago the 11-year cycle could be 'seen' in tree rings. How could we model the Sun spots cycle? Models of a frictionless pendulum are perfectly periodic, while the solar cycle is not. In the 1920s, the Scottish statistician Udny Yule discovered a new model structure, realizing how to introduce randomness into the model and get more realistic-looking time series behaviour. He likened the observed time series of Sun spots to those from the model of a damped pendulum, a pendulum with friction which would have a free period of about 11 years. If this model pendulum were 'left alone in a quiet room', the resulting time series would slowly damp down to nothing. In order to motivate his introduction of random numbers to keep the mathematical model going, Yule extended the analogy with a physical pendulum: 'Unfortunately, boys with pea shooters get into the room, and pelt the pendulum from all sides at random.' The resulting models became a mainstay in the statistician's arsenal. A linear, stochastic mainstay. We will define the **Yule Map**:

Take α times X plus a random value R to be the new value of X where R is randomly drawn from the standard bell-shaped distribution.

So how does this stochastic model differ from a chaotic model? There are two differences that immediately jump out at the mathematician: the first is that Yule's model is stochastic – the rule requires a random number generator, while a chaotic model of the Sun spots would be deterministic by definition. The second is that Yule's model is linear. This implies more than simply that we do not multiply components of the state together in the definition of the map; it also implies that one can combine solutions of the system and get other acceptable solutions, a property called *superposition*. This very useful property is not present in nonlinear systems.

Yule developed a model similar to the Yule Map that behaved more like the time series of real Sun spots. Cycles in Yule's improved model differ slightly from one cycle to the next due to the random effects, the details of the pea shooters. In a chaotic model the state of the Sun differs from one cycle to the next. What about *predictability*? In any chaotic model, almost all nearby initial states will eventually diverge, while in each of Yule's models even far away initial states would converge, *if* both experienced the same forcing from the pea shooters. This is an interesting and rather fundamental difference: similar states diverge under deterministic dynamics whereas they converge under linear stochastic dynamics. That does not necessarily make Yule's model easier to forecast, since we never know the details of the future random forcing, but it changes the way that uncertainty evolves in the system, as shown in

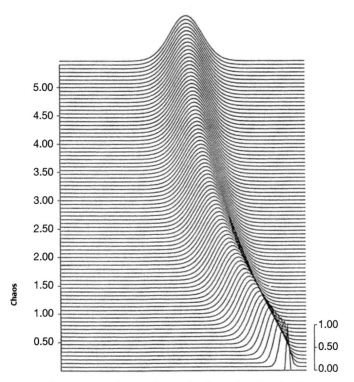

10. The evolution of uncertainty under the stochastic Yule Map. Starting as a point at the bottom of the graph, the uncertainty spreads to the left as we move forward in time (upwards) and approaches a constant bell-shaped distribution

Figure 10. Here an initially small uncertainty, or even an initially zero uncertainty, at the bottom grows wider and moves to the left with each iteration. Note that the uncertainty in the state seems to be approaching a bell-shaped distribution, and has more or less stabilized by the time it reaches the top of the graph. Once the uncertainty saturates in a static state, then all predictability is lost; this final distribution is called the 'climate' of the model.

Physical dynamical systems

There is no way of proving the correctness of the position of 'determinism' or 'indeterminism'. Only if science were complete or demonstrably impossible could we decide such questions.

E. Mach (1905)

There is more to the world than mathematical models. Just about anything we want to measure in the real world, or even just think about observing, can be taken to have come from a physical dynamical system. It could be the position of the planets in the solar system, or the surface of a cup of coffee on a vibrating table, or the population of fish in a lake, or the number of grouse on an estate, or a coin being flipped.

The time series we want to observe now is the state of the physical system: say, the position of our nine planets relative to the Sun, the number of fish or grouse. As a short-hand, we will again denote the state of the system as X, while trying to remember that there is a fundamental difference between a model-state and the True state, if such a thing exists. It is unclear how these concepts stand in relation to each other; as we shall see in Chapter 11, some philosophers have argued that the discovery of chaos implies the real world must have special mathematical properties. Other philosophers, perhaps sometimes the same ones, have argued that the discovery of chaos implies mathematics does not describe the world. Such are philosophers.

In any event, we never have access to the True state of a physical system, even if one exists. What we do have are observations, which we will call 'S' to distinguish them from the state of the system, X. What is the difference between X and S? The unsung hero of science: **noise**. Noise is the glue that bonds the experimentalists with the theorists on those occasions when they meet. Noise is also the grease that allows theories to slide easily over awkward facts.

In the happy situation where we know the mathematical model which generated the observations and we also know of a **noise model** for whatever generated whatever noise there was, then we are in the **Perfect Model Scenario**, or PMS. It is useful to distinguish a strong version of PMS where we know the parameter values exactly, from a weak version where we know only the mathematical forms and must estimate parameter values from the observations. As long as we are in either version of PMS, the noise is defined by the distance between X and S, and it makes sense to speak of noise as causing our uncertainty in the state, since we know a True state exists even if we do not know its value. Not much of this picture survives when we leave PMS. Even within PMS, noise takes on a new prominence once we acknowledge that the world is not linear.

What about the concepts of deterministic and random, or even periodic? These refer to properties of our models; we can apply them to the real world only via (today's) best model. Are there really random physical dynamical systems? Despite the everyday use of coin flips and dice as sources of 'randomness', the typical answer in classical physics is: no, there is no randomness at all. With a complete set of laws it may (or may not) be too difficult for us to calculate the outcomes of coin flips, rolling dice, or spinning a roulette: but that is a problem only in practice, not in principle: Laplace's demon would have no difficulty with such predictions. Quantum mechanics, however, is different. Within the traditional quantum mechanical theory, the half-life of a uranium atom is as natural and real a quantity as the mass of the uranium atom. The fact that classical coin tosses or roulette are not best modelled as random is irrelevant, given the quantum mechanical claim for randomness and objective probabilities. Claims for – or against – the existence of objective probabilities require interpreting physical systems in terms of our models of those systems. As always. Some future theory may revoke this randomness in favour of determinism, but we are on the scene only for a vanishingly small interval. It is relatively safe to say that some of our best models of reality will still admit random elements as you read these words.

Observations and noise

Over the last few decades, a huge number of scientific papers have been written about using a time series to distinguish deterministic systems from stochastic systems. This avalanche was initiated in the physics literature, and then spread into geophysics, economics, medicine, sociology, and beyond. Most of these papers were inspired by a beautiful theorem proven by the Dutch mathematician Floris Takens in 1983, to which we will return in Chapter 8. Why were all these papers written, given that we have a simple rule for determining if a mathematical system is deterministic or stochastic? Why not just look at the rules of the system and see if it requires a random number generator? It is common to confuse the games mathematicians play with constraints placed on the work of the natural (and other) scientists.

Real mathematicians like to play intellectual games, like pretending to forget the rules and then guessing if the system is deterministic or stochastic from looking only at the time series of the states of the system. Could they clearly identify any deterministic system given the time series from the infinitely remote past to the infinitely distant future? For fixed points and even periodic loops, this game is not challenging enough; to make it more interesting, consider a variation in which we do not know the exact states, but have access only to noisy observations, S, of each state X. The origin S is commonly, if somewhat misleadingly, thought of as being related to the addition of a random number to each true X. In that case, this *observational noise* does not affect the future states of the system, only our observations of each state; it is a very different role from that played by the random numbers R in the stochastic systems, like the Yule Map where the value of R did impact the future since it changed the next value of X. To maintain this distinction, random influences that do influence X are called *dynamic noise*.

As noted above, mathematicians can work within the Perfect Model Scenario (PMS). They start off knowing that the model which generated the time series has a certain kind of structure, and sometimes they assume they know the structure (weak PMS), sometimes even the values of the parameters as well (strong PMS). They generate a time series of X, and from this a time series of S. They then pretend to forget the values of X and see if they can work out what they were, or they pretend to forget the mathematical system and see if, given only S, they can identify the system along with its parameter values, or determine if the system is chaotic, or forecast the next value of X.

At this point, it should be pretty easy to see where their game is going: our mathematicians are trying to simulate the situation that natural scientists can never escape from. The physicists, earth scientists, economists, and other scientists do *not* know the rule, the full Laws of Nature, relevant to the physical systems of scientific study. And scientific observations are not perfect; they may be invariably uncertain due to observational noise, but that is not the end of the story. It is a capital mistake to confuse real observations with those of these mathematical games.

The natural scientist is forced to play a different game. While attempting to answer the same questions, the scientist is given only a time series of observations, S, some information regarding the statistics of the observational noise, and the *hope* that some mathematical map exists. Physicists can never be sure if such a structure exists or not; they cannot even be certain if the model state variable X really has any physical meaning. If X is the number of rabbits in a real garden, it is hard to imagine that X does not exist, it is just some whole number. But what about model variables like wind speed or temperature? Are there real numbers that correspond to those components of our state vector? And if not, where between rabbits and wind speed does the correspondence break down?

Our philosopher is very interested in such questions, and we all should be. LeVerrier, the Frenchman who worked with Fitzroy to set up the first weather warning system, died famous for discovering two planets. He used Newton's Laws to predict the location of Neptune based on 'irregularities' in the observed time series of Uranus's orbit, and that planet was duly observed. He also analysed 'irregularities' in the orbit of Mercury, and again told observers where to find another new planet. And they did: the new planet, named Vulcan, was very near the Sun and difficult to see, but it was observed for decades. We now know that there is no planet Vulcan; LeVerrier was misled because Mercury's orbit is poorly described by Newton's Laws (although it is rather better described by Einstein's). How frequently do we blame the mismatch between our models and our data on noise when the root cause is in fact model inadequacy? Most really interesting science is done at the edges, whether the scientists realize it or not. We are never sure if today's laws apply there or not. Modern-day climate science is a good example of hard work being done at the edge of our understanding.

The study of chaos has clarified the importance of distinguishing two different issues: one being the effects of uncertainty in the state or the parameters, the other being the inadequacy of our mathematics itself. Mathematicians working within PMS can make progress by pretending that they are not, while scientists who pretend – or believe – that they are working within PMS when they are not can wreak havoc, especially if their models are naïvely taken as a basis for decision making. The simple fact is that we cannot apply the standards of mathematical proof to physical systems, but only to our mathematical models of physical systems. It is impossible to prove that a physical system is chaotic, or to prove it is periodic. Our physicist and mathematician must not forget that they sometimes use the same words to mean rather different things; when they do, they often run into some difficulty and considerable acrimony. Mach's comment above (page 53) suggests that this is not a new issue.

Chapter 4
Chaos in mathematical models

We would all be better off if more people realised that simple nonlinear systems do not necessarily possess simple dynamical properties.
Lord May (1976)

This chapter consists of a very short survey of chaotic mathematical models from zoology to astronomy. Like any cultural invasion, the arrival of nonlinear deterministic models with sensitive dependence was sometimes embraced, and sometimes not. It has been most uniformly welcomed in physics where, as we shall see, the experimental verification of its prophecies has been nothing short of astounding. In other fields, including population biology, the very relevance of chaos is still questioned. Yet it was population biologists who proposed some of the first chaotic models a decade before the models of astronomers and meteorologists came on the scene. Renewed interest in this work was stimulated in 1976 by an influential and accessible review article in the journal *Nature*. We begin with the basic insights noted in that article.

The darling bugs of May

In 1976, Lord May provided a high-profile review of chaotic dynamics in *Nature* that surveyed the main features of deterministic nonlinear systems. Noting that many interesting questions remained unresolved, he argued that this new perspective provided not just theoretical but practical and pedagogical value as well, and that it suggested everything from new metaphors for describing systems to new quantities to observe and new parameter values to estimate. Some of the simplest population dynamics are those of breeding populations when one generation does not overlap with the next. Insects that have one generation per year, for example, might be described by discrete time maps. In this case X_i would represent the population, or population density, in the ith year, so our time series would have one value per year, and the map is the rule that determines the size of next year's population given this year's. A parameter α represents the density of resources. In the 1950s, Moran and Ricker independently suggested the map shown in Figure 8(f) (page 40). Looking at this graph, we can see that when the current value of X is small, the next value of X is larger: small populations grow. Yet if X gets too big, then the next value of X is small, and when the current value is very large, the next value is very small: large populations exhaust the resources available to each individual, and so successful reproduction is reduced.

Irregularly fluctuating populations have long been observed, and researchers have long argued over their origin. Time series of Canadian lynx and both Scandinavian and Japanese voles are, along with the Sun spot series, some of the most analysed data sets in all of statistics. The idea that very simple nonlinear models can display such irregular fluctuations suggested a new potential mechanism for real population fluctuations, a mechanism that was in conflict with the idea that 'natural' populations should maintain either a steady level or a regular periodic cycle. The idea that these random-*looking* fluctuations need not be induced by some outside force like the weather, but could be inherent to the natural population dynamics, had the potential to radically alter attempts to understand and manage populations. While noting that 'replacing a population's interactions with its biological and physical environment by passive parameters may do great violence to the reality', May provided a survey of interesting behaviours in the Logistic Map. The article ends with 'an evangelical plea for the introduction of these difference equations into elementary mathematics courses, so that students' intuition may be enriched by seeing the wild things that simple nonlinear equations can do'. That was three decades ago.

We will consider a few of these wild things below, but note that the mathematicians' focus on the Logistic Map is not meant to suggest that this map itself in any sense 'governs' the various physical and biological systems. One thing that distinguishes nonlinear dynamics from traditional analysis is that the former tends to focus more on the behaviour of systems rather than on the details of any one initial state under particular equations with specific parameter values: a focus on geometry rather than statistics. Similar dynamics can be more important than 'good' statistics. And it turns out that the Logistic Map and the Moran-Ricker Map are very similar in this way, even though they look very different in Figure 8(f) (page 40). The details may well matter, of course; the enduring role of the Logistic Map itself may be pedagogical, helping to exorcize the historical belief that complicated dynamics requires either very complicated models or randomness.

Universality: prophesying routes to chaos

The Logistic Map gives rise to amazingly rich varieties of behaviour. The famous bifurcation diagram of Figure 11 summarizes the behaviour of the map at many different values of its parameter in one figure. The horizontal axis is α and the dots in any vertical slice indicate states which fall near the attractor for that value of α. Here α reflects some parameter of the system: if X is the number of fish in the lake, then α is the amount of food in the lake; if X is the time between drips of the faucet, then α is the rate of water leaking through the tap; if X is the motion of rolls in fluid convection, then α is the heat delivered to the bottom of the pan. In models of very different things, the behaviour is the same. For small α (on the left) we have a fixed point attractor. The location of the fixed point increases as α increases, until α reaches a value of one, where the fixed point vanishes and we observe iterations which alternate between two points: a period two loop. As α continues to increase, we get a period four loop, then period eight, then 16, then 32. And so on. Bifurcating over, and over again.

Since the period of the loop always increases by a factor of two, these are called *period doubling bifurcations*. While the old loops are no longer seen, they do not cease to exist. They are still there, but have become unstable. This is what happened to the origin in the Logistic Map when α is greater than one: X only stays at zero if it is exactly equal to zero, while small non-zero values grow at each iteration. Similarly, points near an unstable periodic loop move away from it, and so we no longer see them clearly when iterating the map.

There is a regularity hidden in Figure 11. Take any three consecutive values of α at which the period doubles, subtract the first from the second, and then divide that number by the difference between the second and the third. The result leads to the Feigenbaum number, ~4.6692016091. Mitch Feigenbaum discovered these relationships, working with a hand calculator in Los Alamos in the late 1970s, and

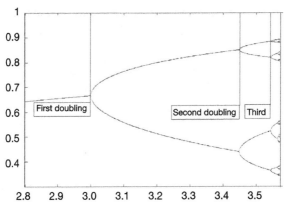

11. Period doubling behaviour in the Logistic Map as α increases from 2.8 to ~3.5; the first three doublings are marked

the ratio is now known by his name. Others also found it independently; having the insight to do this calculation was stunning in each case.

Since the Feigenbaum number is greater than one, values of α at which bifurcations occur get closer and closer together, and we have an infinite number of birfurcations before reaching a value of α near 3.5699456718. Figure 12 indicates what happens for larger values of α. This sea of points is largely chaotic. But note the windows of periodic behaviour, for instance the period three window where α takes on the value of one plus the square root of eight (that is, about 3.828). This is a stable period three loop; can you identify windows corresponding to period five? Seven?

Figure 13 puts the figures of the Logistic Map in context. Randomly chosen values for α and X_0 form a cloud of points on the t equals zero slice of this three-dimensional figure. Iterating the Logistic Map forward from these values, the transients fall away, and the attractors at each value of α slowly come into view, so that after 512 iterations the last time slice resembles Figure 12.

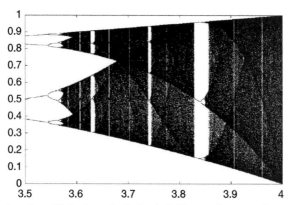

12. **A variety of behaviours in the Logistic Map as α increases from a period four loop at α = 3.5 to chaos at α = 4. Note the replicated period doubling cascades at the right side of each periodic window**

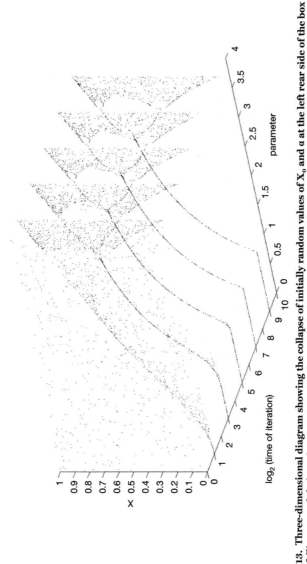

13. Three-dimensional diagram showing the collapse of initially random values of X_0 and α at the left rear side of the box falling toward their various attractors as the number of iterations increases. Note the similarity of the points near the right forward side with those in Figures 11 and 12

It would be asking too much to expect something as simple as the Logistic Map to tell us anything about the behaviour of liquid helium. But it does. Not only does the onset of complicated behaviour show a qualitative indication of period doubling, the actual quantitative values of the Feigenbaum numbers computed from many experiments agree remarkably well with those computed from the Logistic Map. Many physical systems seem to display this 'period doubling route to chaos': hydrodynamics (water, mercury, and liquid helium), lasers, electronics (diodes, transistors), and chemical reactions (BZ reaction). One can often estimate the Feigenbaum number to two-digits' accuracy in experiments. This is one of the most astounding results reported in this Introduction to chaos: how could it be that simple calculations with the Logistic Map can give us information that is relevant to all these physical systems?

The mathematician's fascination with this diagram arises not only from its beauty but also from the fact that we would get a similar picture for the Moran-Ricker Map and many other systems that at first instance appear quite different from the Logistic Map. A technical argument shows that the period doubling is common in 'one-hump' maps where the hump *looks like* a parabola. In a very real and relevant sense, almost all nonlinear maps look like this very close to their maximum value, so properties like period doubling are called 'universal', although not *all* maps have them. More impressive than these mathematical facts is the empirical fact that a wide variety of physical systems display unexpected behaviour that, as far as we can see, reflects this mathematical structure. Is this not then a strong argument for the mathematics to govern, not merely describe, Nature? To address this question, we might consider whether the Feigenbaum number is more akin to a constant of geometry, like π, or a physical constant like the speed of light, c. The geometry of disks, cans, and balls is well described using π, but π hardly governs the relationship between real lengths, area, and volumes in the same way that the values of physical constants govern the nature of things within our laws of nature.

The origin of the mathematical term 'chaos'

In 1964 the Russian mathematician A. N. Sharkovski proved a remarkable theorem about the behaviours of many 'one-hump' maps: namely that discovering a periodic loop indicated that others, potentially lots of others, existed. Discovering that a period 16 loop existed for a particular value of the parameter implied there were loops of period eight and of four and of two and of one at that value; while finding a loop of period three meant that there was a loop of every possible period! It is another non-constructive proof; it does not tell us where those loops are, but nevertheless it is a pretty neat result. Eleven years after Sharkovski, Li and Yorke published their enormously influential paper with the wonderful title 'Period Three Implies Chaos'. The name 'chaos' stuck.

Higher-dimensional mathematical systems

Most of our model states so far have consisted of just one component. The vole and weasel model is an exception, since the state consisted of two numbers: one reflecting the population of voles, the other the population of weasels. In this case, the state is a vector. Mathematicians call the number of components in the state the *dimension* of the system, since plotting the state vectors would require a state space of that dimension.

As we move to higher dimensions, the systems are often not maps but *flows*: a map is a function that takes one value of X and returns the next value of X, while a flow provides the velocity of X for any point in the state space. Think of a parsnip floating under the surface of the sea; it is carried along by the current and will trace out the flow of the sea. The three-dimensional path of the parsnip in the sea is analogous to a path traced out by X in the state space, and both are sometimes called *trajectories*. If instead of a parsnip, we follow the path of an infinitesimal parcel of the fluid itself, we often find these paths to be recurrent with sensitive dependence. The equations are deterministic and these fluid parcels are said to display 'Lagrangian chaos'. Laboratory experiments with fluids often display beautiful patterns which reflect the chaotic dynamics observed in our models of fluid flow. Without examining the differential equations that define these velocity fields, we will next touch several classic chaotic systems.

Dissipative chaos

In 1963, Ed Lorenz published what became a classic paper on the predictability of chaotic systems. He considered a vastly simplified set of three equations based on the dynamics of a fluid near the onset of convection which is now called the *Lorenz System*. One can picture the three components of the state in terms of convective rolls in a layer of fluid between two flat plates when the lower plate is heated. When there is no convection, the fluid is motionless and the temperature in the fluid decreases uniformly from the warmer plate at the bottom to the cooler plate at the top. The state X of the Lorenz model consisted of three values {x,y,z}, where x reflected the speed of the rotating fluid, y quantified the temperature difference between rising and sinking fluid, and z measured the deviation from the linear temperature profile. An attractor from this system is shown in Figure 14; by chance, it looks something like a butterfly. The different shading on the attractor indicates variations in the time it takes an infinitesimal uncertainty to double. We return to discuss the meaning of these shades in Chapter 6, but note the variations with location.

The evolution of uncertainty in the Lorenz system is shown in Figure 15; this looks a bit more complicated than the corresponding figure for the Yule Map in Figure 10 (page 52). Figure 15 shows the kind of forecast our 21st-century demon could make for this system: an initial small uncertainty at the bottom of the panel grows wider, then narrower, then wider, then narrower . . . eventually splitting into two parts and beginning to fade away. But depending on the decisions we are trying to make, there may still be useful information in this pattern even at the time reflected at the top of

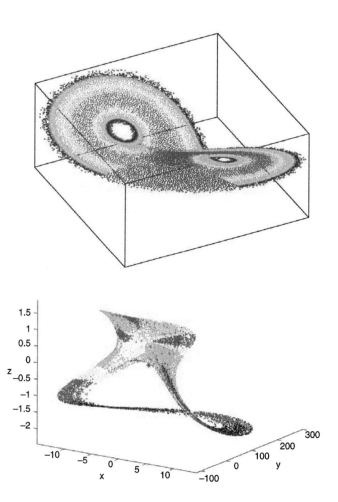

14. Three-dimensional plots of (above) the Lorenz attractor and (below) the Moore-Spiegel attractor. The shading indicates variations in uncertainty doubling time at each point

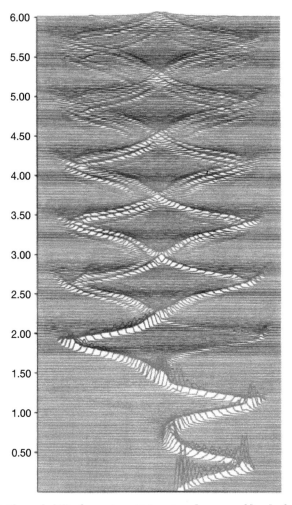

15. The probability forecast our 21st-century demon would make for the 1963 Lorenz System. Contrast the way uncertainty evolves in this chaotic system with the relatively simple growth of uncertainty under the Yule Map shown in Figure 10 on page 52

the panel. On this occasion the uncertainty has not stabilized by the time it reaches the top of the graph.

In 1965, mathematical astronomers Moore and Spiegel considered a simple model of a parcel of gas in the atmosphere of a star. The state space is again three-dimensional, and the three components of X are simply the height, velocity, and acceleration of the parcel. The dynamics are interesting because we have two competing forces: a thermal force that tends to destabilize the parcel and a magnetic force that tends to bring it back to its starting point, much like a spring would. As the parcel rises, it finds itself at a different temperature than the surrounding fluid and this feeds back on its velocity and its temperature, but at the same time the star's magnetic field acts as a spring to pull the parcel back towards its original location. Motion caused by two competing forces often gives rise to chaos. The Moore-Spiegel attractor is also shown in Figure 16.

Chaos experiments have always pushed computers to their limits, and sometimes slightly beyond those limits. In the 1970s, the astronomer Michael Hénon wanted to make a detailed study of chaotic attractors. For a given amount of computer power there is a direct trade-off between the complexity of the system and the duration of the time series one can afford to compute. Hénon wanted a system with properties similar to Lorenz's 1963 system that would be cheaper to iterate on his computer. This was a two-dimensional system, where the state X consisted of the pair of values {x,y}. The Hénon Map is defined by the rules:

The new value of x_{i+1} is equal to one minus y_i plus α times x_i^2; the new value of y_{i+1} is equal to β times x_i.

Panel (b) of Figure 16 shows the attractor when α is 1.4 and β is 0.3; panel (a) shows a slice of the Moore-Spiegel attractor made by combining snapshots of the system whenever z was zero and

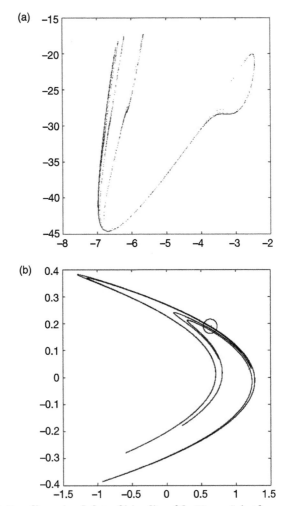

16. **Two-dimensional plots of (a) a slice of the Moore-Spiegel attractor at z = 0; and (b) the Hénon attractor where α is 1.4 and β is 0.3. Note the similar structure with many leaves in each case**

growing. This type of figure is called a ***Poincaré section*** and illustrates how slices of a flow are much like maps.

Delay equations, epidemics, and medical diagnostics

Another interesting family of models are delay equations. Here both the current state and some state in the past (the 'delayed state') play a direct role in the dynamics. These models are common for biological systems, and can provide insight into oscillatory diseases like leukaemia. In the blood supply, the number of cells available tomorrow depends upon the number available today, and also the number of new cells that mature today; the delay comes from the gap in time between when these new cells are requested and when they mature: the number of cells maturing today depends on the number of blood cells at some point in the past. There are many other diseases with this kind of oscillatory dynamics, and the study of chaos in delay equations is extremely interesting and productive.

We leave the discussion of mathematical models for a paragraph to note that medical research is another area where insights from our mathematical models are deployed for use in real systems. Research by Mike Mackey at McGill University and others on delay equations has even led to a cure for at least one oscillatory disease. The study of nonlinear dynamics has also led to insights in the evolution of diseases that oscillate in a population, not an individual; our models can be contrasted with reality in the study of measles, where one can profitably consider the dynamics in time and in space. The analysis of chaotic time series has also led to the development of insightful ways to view complicated medical time series, including those from the brain (EEG) and heart (ECG). This is not to suggest that these medical phenomena of the real word are chaotic, or even best described with chaotic models; methods of analysis developed for chaos may prove of value in practice regardless of the nature of the underlying dynamics of the real systems that generate the signals analysed.

Hamiltonian chaos

If volumes in state space do not shrink in time there can be no
attractors. In 1964, Hénon and Heiles published a paper showing
chaotic dynamics in a four-dimensional model of the motion of a
star within a galaxy. Systems in which volumes in state space do not
shrink, including those of Newtonian celestial mechanics commonly
used to predict eclipses, and which trace the future of the solar
system and spacecraft within it, are called *Hamiltonian*. Figure 17
is a slice from the Hénon-Heiles system, which is Hamiltonian.
Note the intricate interweaving of empty islands in a sea of chaotic
trajectories. Initial states started within these islands may fall onto
almost closed loops (tori); alternatively they may follow chaotic
trajectories confined within an island chain. In both cases, the order
in which the islands in the chain are visited is predictable, although
exactly where on each island might be unpredictable; in any case,
things are only unpredictable on small length scales.

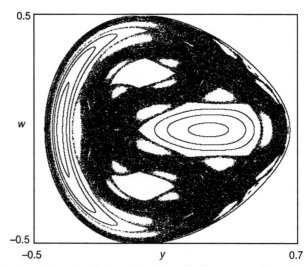

**17. A two-dimensional slice of the Hénon-Heiles attractor. Note the
simultaneous loops, and a chaotic sea with many (empty) islands**

Exploiting the insights of chaos

In the three-year period between 1963 and 1965, three independent papers appeared (by Lorenz, by Moore and Spiegel, and by Hénon and Heiles), each using digital computers to introduce what would be called 'chaotic dynamics'. In Japan, chaos had been observed in analogue computer experiments by Yoshisuke Ueda, and Russian mathematicians were advancing upon a groundwork laid down by over a century of international mathematics. Almost 50 years later, we are still finding new ways to exploit these insights.

What limits the predictability of future solar eclipses? Is it uncertainty in our knowledge of the planetary orbits due to the limited accuracy of our current measurements? Or future variations in the length of the day which alters the point on the surface of the Earth under the eclipse? Or the failure of Newton's equations due to effects (better) described by general relativity? We can see that the Moon is slowly moving away from the Earth, and assuming that this continues, it will eventually appear too small to block the entire Sun. In that case, there will be a last total eclipse of the Sun. Can we forecast when that event will occur and, weather permitting, where we should be on the surface of the Earth in order to see it? We do not know the answer to that question. Nor do we know, for certain, if the solar system is stable. Newton was well aware of the difficulties nonlinearities posed for determining the ultimate stability of only three celestial bodies, and suggested that insuring the stability of the solar system was a task for God. By understanding the kinds of chaotic orbits that Hamiltonian systems admit, we have learned many things about the ultimate stability of the solar system. Our best guess, currently, is that our solar system is stable, probably. Insights like these come from understanding the geometry in state space rather than attempting detailed calculations based upon observations.

Can we safely draw insights from mathematical behaviour of low-dimensional systems? They suggest new phenomena to look for in experiments, like periodic doubling, or suggest new constants to estimate in Nature, like the Feigenbaum number. These simple systems also provide test beds for our forecast methods; this is a bit more dangerous. Are the phenomena of low-dimensional chaotic systems the same phenomena that we observe in more complicated models? Are they so common that they occur *even in* simple low-dimensional systems like Lorenz 1963 or the Moore-Spiegel system? Or are these phenomena due to the simplicity of these examples: do they occur *only in* simple mathematical systems? The same *even in or only in* question applies to techniques developed to forecast or control chaotic systems, which are tested in low-dimensional systems: do these things happen *even in or only in* low-dimensional systems? The most robust answer so far is that difficulties we identify in low-dimensional systems rarely go away in higher-dimensional systems, while successful solutions to these difficulties which work in low-dimensional systems often fail to work in higher-dimensional systems. Recognizing the danger of generalizing from three-dimensional systems, Lorenz moved on to a 28-dimensional system about 50 years ago; he is still creating new systems today, some in two dimensions and others in 200 dimensions.

Chaos and nonlinearity impact many fields; perhaps the deepest insight to be drawn here is that complicated-looking solutions are sometimes acceptable and need not be due to external dynamic noise. This does not imply that, in any particular case, they are not due to external noise, nor does it lessen the practical value of stochastic statistical modelling, which has almost a century of experience and statistical good practice behind it. It does suggest the value in developing tests for which methods to use in a given application, and consistency tests for all modelling approached. Our models should be as uninhibited as possible, but not more so. The lasting impact of these simple systems may be in their pedagogical value; young people can be exposed to the rich behaviours of these simple systems early in their education. By requiring internal consistency, mathematics constrains our flights of fancy in drawing metaphors, not so much as to bring them in line with physical reality, but often opening new doors.

Chapter 5
Fractals, strange attractors, and dimension(s)

Big fleas have little fleas
upon their backs to bite'em.
And little fleas have lesser fleas,
and so ad infinitum.

      A. de Morgan (1872)

No introduction to chaos would be complete without touching upon **_fractals_**. This is neither because chaos implies fractals nor because fractals require chaos, but simply because in dissipative chaos real mathematical fractals appear as if from nowhere. It is just as important to distinguish mathematical fractals from physical fractals as it is to distinguish what we mean by chaos in mathematical systems from what we mean by chaos in physical systems. Despite decades of discussion, there is no single generally accepted definition of a fractal in either case, although you can usually recognize one when you see it. The notion is bound up in self-similarity: as we zoom in on the boundary of clouds, countries, or coastlines, we see patterns similar to those seen at the larger-length scales again and again. The same thing happens with the set of points in Figure 18. Here the set is composed of five clusters of points; if we enlarge any one of these clusters, we find the enlargement looks similar to the entire set itself. If this similarity is exact – if the zoom is equivalent to the original set – then the set is called _strictly self-similar_. If only statistical properties of interest are repeated, then the set is called *statistically self-similar*. Deciding exactly what counts as a 'statistical property of interest' opens one of the discussions that has prevented agreement on a general definition. Disentangling these interesting details deserves its own *Very Short Introduction to Fractals*; we will content ourselves with some examples.

In the late 18th century, fractals were widely discussed by mathematicians including Georg Cantor, although the famous Middle Thirds set that bears his name was first found by an Oxford mathematician named Henry Smith. Fractal entities were often disavowed by their mathematical parents as monstrous curves in the 100 years that followed, just as L. F. Richardson was beginning to quantify the fractal nature of various physical fractals. Both physical and mathematical fractals were more warmly embraced by astronomers, meteorologists, and social scientists. One of the first fractals to bridge the divide – and blur the distinction – between a mathematical space and real-world space appeared about 100 years ago in an attempt to resolve Olbers' paradox.

A fractal solution to Olbers' paradox

In 1823, the German astronomer Heinrich Olbers encapsulated a centuries-old concern of astronomers in the concise question: 'Why is the night sky dark?' If the universe were infinitely large and more or less uniformly filled with stars, then there would be a balance between the number of stars at a given distance and the light we get from each one of them. This delicate balance implies that the night sky should be uniformly bright; it would even be difficult to see the Sun against a similarly bright day-time sky. But the night sky is dark. That is Olbers' paradox. Johannes Kepler used this as an argument for a finite number of stars in 1610. Edgar Allan Poe was the first to suggest an argument still in vogue today: that the night sky was dark because there had not been enough time for light from far-away stars to reach the Earth, yet. Writing in 1907, Fournier d'Albe proposed an elegant alternative, suggesting that the distribution of matter in the universe was uniform but in a fractal manner. Fournier illustrated his proposal with the figure reproduced in Figure 18. This set is called the Fournier Universe. It is strictly self-similar: blowing up one of the small cubes by a factor of 5 yields an exact duplicate of the original set. Each small cube contains the totality of the whole.

The Fournier Universe illustrates a way out of Olbers' paradox: the line Fournier placed in Figure 18 indicates one of many directions in which no other 'star' will ever be found. Fournier did not stop

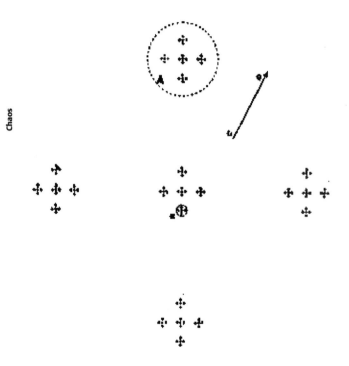

DIAGRAM OF A MULTI-UNIVERSE

18. The Fournier Universe, showing the self-similar structure, as published by Fournier himself in 1907

at the infinitely large, but also suggested that this cascade actually extended to the infinitely small; he interpreted atoms as micro-verses, which were in turn made of yet smaller particles, and suggested macro-verses in which our galaxies would play the role of atoms. In this way, he proposed one of the few physical fractals with no inner cut-off and no outer cut-off: a cascade that went from the infinitely large to the infinitesimally small in a manner reminiscent of the last frames of the film *Men in Black*.

Fractals in physics

Big whorls have little whirls,
which feed on their velocity.
And little whirls have lesser whirls,
and so on to viscosity.

L. F. Richardson

Clouds, mountains, and coastlines are common examples of natural fractals; statistically self-similar objects that exist in real space. Interest in generating fractal irregularity is not new: Newton himself recorded an early recipe, noting that when beer is poured into milk and 'the mixture let stand till dry, the surface of the curdled substance will appear as rugged and mountainous as the earth at any place'. Unlike Newton's curdled substance, the fractals of chaos are mathematical objects found in state spaces; they are true fractals as opposed to their physical counterparts. What is the difference? Well, for one thing, a physical fractal only displays the properties of a fractal at certain length scales and not at others. Consider the edge of a cloud: as you look more and more closely, going to smaller and smaller length scales, you'll reach a point at which the boundary is no more; the cloud vanishes into the helter-skelter rush of molecules and there is no boundary to measure. Similarly, a cloud is not self-similar on length scales comparable with the size of the Earth. For physical fractals, fractal concepts break down as we look too closely; these physical cut-offs make it easy to identify old Hollywood special effects using model ships in wave tanks: we can sense the cut-off is at the wrong length scale relative to the 'ships'. Today, film makers in Hollywood and in Wellington have learned enough mathematics to generate computer counterfeits that hide the cut-off better. The Japanese artist Hokusai respected this cut-off in his famous 'Great Wave' print of the 1830s. Physicists have also known this for some time: de Morgan's poem allowed its cascade of fleas to continue *ad infinitum*, while the cascade whorls in L. F. Richardson's version face a limit due to viscosity, the term for friction within fluids. Richardson was expert in the theory and observation of turbulence. He once threw parsnips off one end of Cape Cod canal at regular intervals, using the time of their arrival at a bridge on the other end of the canal to quantify how the fluid dispersed as it moved downstream. He also computed (by hand!) the first numerical weather forecast, during the First World War.

A Quaker, who left the Met Office in the First World War to become an ambulance driver in France, Richardson later became interested in measuring the length of the border between nations in order to test his theory that this influenced the likelihood of their going to war. He identified an odd effect when measuring the same border on different maps: the border between Spain and Portugal was much longer when measured on the map of Portugal than it was when measured on the map of Spain! Measuring coastlines of island nations like Britain, he found that the length of the coastline increased as the callipers he walked along the coast to measure it decreased, and also noted an unexpected relationship between the area of an island and its perimeter as both vary when measured on different scales. Richardson demonstrated that these variations with length scale followed a very regular pattern which could be summarized by a single number for a particular boundary: an exponent that related the length of a curve to the length scale used to measure it. Following fundamental work by Mandelbrot, this number is called the *fractal dimension* of the boundary.

Richardson developed a variety of methods to estimate the fractal dimension of physical fractals. The area-perimeter method quantifies how the area and perimeter both change under higher and higher resolution. For one particular object, such as a single cloud, this relationship also yields the fractal dimension of its border. When we look at many *different* clouds at the *same* resolution, as in a photograph from space, a similar relationship between areas and perimeters emerges; we do not understand why this alternative area-perimeter relation seems to hold for collections of different-sized clouds, given that clouds are famous for not all looking the same.

Fractals in state space

We next construct a rather artificial mathematical system designed to dispel one of the most resilient and misleading myths of chaos: that detecting upon a fractal set in state space indicates deterministic dynamics. The Tripling Tent Map is:

> If X is less than a half then take 3X as the new value of X,
> Otherwise take 3 minus 3X as the new value of X.

Almost every initial state between zero and one flies far away from the origin; we will ignore these and focus on the infinite number of initial conditions which remain forever between zero and one. (We ignore the apparent paradox due to the loose use of 'infinity' here, but note Newton's warning that 'the principle that all infinities are equal is a precarious one'.)

The Tripling Tent Map is chaotic: it is clearly deterministic, the trajectories of interest are recurrent, and the separation between infinitesimally close points increases by a factor of three on each iteration, which implies sensitive dependence. A time series from the Tripling Tent Map, along with one from the stochastic Middle Thirds IFS Map, are shown in Figure 19. Visually, we see hints that

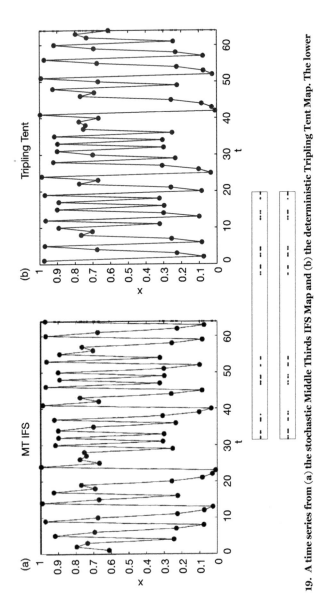

19. A time series from (a) the stochastic Middle Thirds IFS Map and (b) the deterministic Tripling Tent Map. The lower insets show a summary of all the points visited: approximations to the Middle Thirds Cantor set in each case

the chaotic map is easier to forecast: small values of X are *always* followed by small values of X The two small insets at the bottom of Figure 19 each show a set of points visited by a long trajectory from one of the systems, they look very similar and in fact both reflect points from the Middle Thirds Cantor set. The two dynamical systems each visit the same fractal set, so we can never distinguish the deterministic system from the stochastic system if we only look at the dimension of the set of points each system visits; but is it any surprise that to understand the dynamics we have to examine how the system moves about, not only where it has been? This simple counter-example slays the myth noted above; while chaotic systems may often move on fractal sets, detecting a finite dimensional set indicates neither determinism nor chaotic dynamics.

Finding fractals in carefully crafted mathematical maps is not so surprising, as mathematicians are clever enough to design maps which create fractals. One of the neatest things about dissipative chaos is that fractals appear without the benefit of intelligent design. The Hénon Map is the classic example. Mathematically speaking, it represents an entire class of interesting models; there is nothing particularly 'fractal-looking' in its definition, as there is in the Middle Thirds IFS Map. Figure 20 shows a series of zooms from where, as if by magic, self-similar structures spring out. Surely this is one of the most amazing things about nonlinear dynamical systems. There is no hint of artificial design in the Hénon Map, and fractal structure appears commonplace in the attractors of dissipative chaotic systems. It is not required for chaos, nor vice versa, but it is common.

Like all magic, we can understand how the trick works, at least after the fact: we have chosen to zoom in about a fixed point of the Hénon Map, and looking at the properties of the map very, very close to this point reveals how much to zoom in order to make its self-similarity so striking. The details of the repeated structure, a thick line and two thinner lines, depend on what happens far away from this point. But if Hénon is really chaotic and the computer

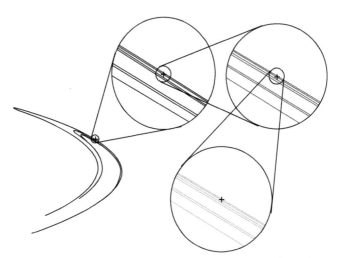

20. A series of zooms into the unstable fixed point of the Hénon Map, which is marked with a ' + ' on each zoom. The same pattern repeats over and over, until we start running out of data points

trajectory used to make these pictures is realistic, then we have a fractal attractor naturally.

The traditional theory of turbulence in state space reflected Richardson's poem: it was thought that more and more periodic modes would be excited and tracing the linear sum of all those oscillations would require a very high-dimensional state space. So most physicists were expecting the attractors of turbulence to be high-dimensional doughnuts, or mathematically speaking, tori. In the early 1970s, David Ruelle and Floris Takens were looking for alternatives to smooth high-dimensional tori and ran into lower-dimensional fractal attractors; they found the fractal attractors 'strange'. Today, the word 'strange' is used to describe the geometry of the attractor, specifically the fact that it is a fractal, while the word 'chaos' is used to describe the dynamics of the system. It is a useful distinction. The precise origin of the phrase **'*strange attractor*'** has been lost, but the term has proven an inspiring and appropriate label for these objects of mathematical physics. Since Hamiltonian systems have no attractors at all, they have no strange attractors. Nevertheless, chaotic time series from Hamiltonian systems often develop intricate patterns with stark inhomogeneity and hints of self-similarity called *strange accumulators* which persist for as long as we run our computers. Their ultimate fate remains unknown.

Fractal dimensions

Counting the number of components in the state vector tells us the dimension of the state space. But how would we estimate the dimension of a set of points if those points do not define a boundary; the points that form a strange attractor, for example? One approach reminiscent of the area-perimeter relation is to completely cover the set with boxes of a given size, and see how the number of boxes required increases as the size of the individual boxes gets smaller. Another approach considers how the number of points changes, on average, as you look inside a ball centred on a random point and decrease the radius of the ball. To avoid complications that arise near the edge of an attractor, our mathematician will consider only balls with a vanishingly small radius, r. We find familiar-looking results: near a random point on a line the number of points is proportional to r^1, about a point in a plane it is proportional to πr^2, and about a point from the set which defines a solid cube, it is proportional to $4/3\, \pi r^3$. In each case, the exponent of r reflects the dimension of the set: one if the set forms a line, two if a plane, three if a solid.

This method can be applied to fractal sets, although fractals tend to have holes, called lacunae, on all scales. While dealing with these logarithmic wrinkles is non-trivial, we can compute the dimension of strictly self-similar sets exactly, and immediately notice that the dimension of a fractal is often not a whole number. For the Fournier Universe, the dimension is ~0.7325 (it equals log 5/log 9) while the Middle Thirds Cantor set has dimension ~0.6309 (it equals log 2/log 3); in each case, the dimension is a fraction bigger than zero yet less than one. Mandelbrot took the 'fract' in 'fraction' as the root of the word 'fractal'.

What is the dimension of the Hénon attractor? Our best estimate is ~1.26, but while we know there is an attractor, we do not know for certain whether or not, in the long run, this attractor is merely a long periodic loop. In maps, every periodic loop consists of only a finite number of points and so has dimension zero. To see this, just consider balls with a radius r smaller than the closest pair of points on the loop; the number of points in each ball is constant (and equal to one), which we can write as proportional to r^0, and so each has dimension zero. In Chapter 7, we shall see why it is hard to prove what happens in the long run using a computer simulation. First, we will take a closer look at the challenges to quantifying the dynamics of uncertainty even when we know the mathematical system perfectly. For real-world systems, we only have noisy observations, and the problem is harder still.

Chapter 6
Quantifying the dynamics of uncertainty

Chaos exposes our prejudices when we examine the dynamics of uncertainty. Despite the hype regarding unpredictability, we shall see that the quantities used to establish chaos place no restriction whatsoever on the accuracy of today's forecast: chaos does not imply that prediction is hopeless. We can see why the link between chaos and predictability has been so badly overstated by looking at the history of the statistics used to measure uncertainty. Additional statistics are available today.

Once scientists touch on uncertainty and predictability, they are honour-bound to clarify the relevance of their forecasts and the statistics used to quantify their uncertainty. The older man looking out of la Tour's painting may have provided the younger man with accurate tables of probabilities for every hand from a deck of 52 cards, but he knows those probabilities do not reflect the game being played. Likewise, our 21st-century demon can quantify the dynamics of uncertainty quite accurately, given her perfect model, but we know we do not have a perfect model. Given only a collection of imperfect models, how might we relate the diversity of their behaviours to our uncertainty about the future state of the real world?

The decay of certainty: information without correlation

When it comes to predicting what a system will do next, data on the recent state of the system often provide more information than data on some long past state of the system. In the 1920s, Yule wanted to quantify the extent to which data on this year's Sun spots provide more information about the number of spots that will appear next year than ten-year-old data do. Such a statistic would also allow him to quantitatively compare properties of the original data with those of time series generated by models. He invented what is now called the auto-correlation function (or ACF), which measures the linear correlation between states k iterations apart. When k is zero the ACF is one, since any number is perfectly correlated with itself. If the time series reflects a periodic cycle, the ACF decreases from one as k increases, and then returns to equal one whenever k is an exact multiple of the period. Given data from a linear stochastic system the ACF is of great value, but as we will soon see, it is of less use when faced with observations from a nonlinear system. Nevertheless, some statisticians went so far as to define determinism as linear correlation; many are still reeling from this misstep. It is well known that correlation does not imply causation; the study of chaos has made it clear that causation does not imply (linear) correlation either. The correlation between consecutive states of the Full Logistic Map is zero despite the fact that the next state is completely determined by the current state. In fact, its ACF is zero for every separation in time. How then are we to detect relationships in nonlinear systems, much less quantify predictability, if a mainstay of a century of statistical analysis is blind to such visible relationships? To answer this question, we first introduce base two.

Bits and pieces of information

Computers tend to record numbers in binary notation: rather than use the ten symbols (0,1,2,3,4,5,6,7,8, and 9) we learn in school, they use only the first two (0 and 1). Instead of 1000, 100 and 10 representing 10^3, 10^2 and 10^1, in binary these symbols represent 2^3, 2^2 and 2^1 that is, eight, four, and two. The symbol 11 in base two represents $2^1 + 2^0$, i.e. three, while 0.10 represents 2^{-1} (one-half) and 0.001 represents 2^{-3} (one-eighth). Hence the joke that there are ten kinds of mathematicians in the world: those who understand binary notation and those who do not. Just as multiplying by ten (10) is easy in base ten, multiplying by two (10) is easy in base two: just shift all the bits to the left, so that 1.0100101011 becomes 10.100101011, that is where the Shift Map gets its name. Similarly dividing by two: its just a shift to the right.

A computer usually uses a fixed number of bits for each number, and does not waste valuable memory space storing the 'decimal' point. This makes dividing a bit curious: On a computer, dividing 001010010101100 by two yields 000101001010110; but then dividing 001010010101101 by two yields the same result! Multiplying 000101001010110 by two yields 00101001010110Q, where Q is a new bit the computer has to make up. So it is for every shift left: a new bit is required in the empty place on the far right. In dividing by two, a zero correctly appears in the empty place on the far left, but any bits that are shifted out the right side this window are lost forever into the bit bucket. This introduces an annoying feature: if we take a number and divide by two, and then multiply by two, we may not get back to the original number we started with.

The discussion thus far leads to differing visions of the growth and decay of uncertainty – or creation of information – in our various kinds of mathematical dynamical systems: random systems, chaotic mathematical systems and computerized versions of chaotic mathematical systems. The evolution of the state of a system is often visualized as a tape passing through a black-box. What happens inside the box depends on what kind of dynamical system we are watching. As the tape exits the box we see the bits written on it; the question of whether the tape is blank when it enters the back of the box, or if it already has the bits written on it, leads to spirited discussions in ivory tower coffee rooms. What are the options? If the dynamics are random, then the tape comes into the box blank and leaves with a randomly determined bit stamped on it. In this case, any pattern we believe we see in the bits as the tape ticks constantly forward is a mirage. If the dynamical system is deterministic, the bits are already printed on the tape (and unlike us, Laplace's demon is in a position to already see all of them); we cannot see them clearly until they pass through the box, but they are already there. Creating all those bits of information is something like a miracle either way, and it seems to come down to personal preference whether you prefer one big miracle or a regular stream of small ones: in a deterministic system the picture corresponds to creating an infinite number of bits all at once: the irrational number which is the initial state; in the random system, it looks as if new bits are created as at each iteration. In practice, it certainly seems that we do have some control over how accurately we measure something, suggesting that the tape is pre-printed.

There is nothing in the definition of a chaotic system that prevents the tape from running backwards for a while. When this happens, prediction gets simple for a while, since we have seen the tape back up, we already know the next bits that will come out when it runs forward again. When we try to cast this image into the form of a computational system, we run into difficulty. The tape cannot really be blank before it comes into the box: the computer has to 'make up' those new bits with some deterministic rule when it left-shifts, so they are effectively already printed on the tape before it enters the box. More interesting is what happens in a region where the tape backs up, since the computer cannot 'remember' any bits it loses on a right-shift. For constant slope maps we are always shifting left or always right, the tape never backs up. The computer simulation is still a deterministic system, although the variety of tapes it can produce is much less rich than the tapes of the deterministic mathematical map it is simulating. If the map being simulated has regions of shrinking uncertainty, then there is a transient period during which the tape backs up, the computer cannot know which bits were written on it; when the tape runs forward again the computer uses its internal rule to make up new bits and we may find a 0 and a 1 overprinted on the tape as it comes out of the box a second time! We discuss other weird things that happen in computer simulations of chaotic mathematical systems in Chapter 7.

Statistics for predicting predictability

One of the insights of chaos is to focus on information content. In linear systems variance reflects information content. Information content is more subtle in nonlinear systems, where size is not the only indicator of importance. How else might we measure information? Consider the points on a circle on the X,Y plane with a radius equal to one, and pick an angle at random. Knowing the value of X tells us a great deal about the value of Y – it tells us that Y is one of two values. Likewise, if we do not know all of the bits needed to completely represent X, the more bits of X we learn, the more bits of Y we know. Although we will never be able to decide between two alternative locations of Y, our uncertainty regarding the two possible locations shrinks as we measure X more and more accurately. Not surprisingly, X and Y have a linear correlation of zero in this case. Other statistical measures have been developed to quantify just how much knowing one value tells you about the other. *Mutual Information*, for instance, reflects how many bits of Y you learn, on average, when you learn another bit of X. For the circle, if you know the first five bits of X, you know four of the first five bits of Y; if you know 20 bits of X, you know 19 of Y; and if you know all the bits of X, you know all but one of the bits of Y. Without that missing bit, we can't tell which of two possible values of Y is the actual value of Y. And unfortunately, from the linear-thinking point of view, the bit you are missing is the value of the 'largest' bit in Y. Nevertheless, it is more than a bit misleading to interpret the fact that the correlation is zero to mean you learn nothing about Y upon learning the value of X.

What does Mutual Information tell us about the dynamics of the Logistic map? Mutual Information will reflect the fact that knowing one value of X exactly gives us complete information on future values of X. While given a finite precision measurement of X, Mutual Information reflects how much we know, on average, about a future measurement of X. In the presence of observational noise we would tend to know less about future values of X the further they fall in the future since the corresponding bits of the current value of X will be obscured by the noise. So Mutual Information tends to decay as the separation in time increases, while the linear correlation coefficient is zero for all separations (except zero). Mutual Information is one useful tool; the development of custom-made statistics to use in particular applications is a growth industry within nonlinear dynamics. It is important to know exactly what these new statistics are telling us, and it is equally important to accept that there is more to say than traditional statistics can tell us.

Our model of the noise gives us an idea of our current uncertainty, so one measure of predictability would be the time we expect that uncertainty to double. We must avoid the trap of linear thinking that suggests the quadrupling time will be twice the doubling time in a non-linear system. Since we do not know which time will be of interest (the doubling-time, tripling-time, quadrupling time, or . . .), we will simply refer to the q-tupling time near a particular initial condition. The distribution of these q-tupling times is relevant to predictability: they directly reflect the time we expect our uncertainty in each particular forecast to go through a given threshold of interest to us. The average uncertainty doubling time gives the same information averaged over forecasts from this model. It is convenient to have a single number, but this average may not apply to any initial state at all.

The average uncertainty doubling time is a useful statistic of predictability. But the definition of mathematical chaos is not made in relation to doubling (or any q-tupling) time statistic, but rather in relation to **Lyapunov exponents** which we define below. This is one reason that chaos and predictability are not as closely related as they are commonly thought to be. The average doubling time gives a more practical indication of predictability than the leading Lyapunov exponent, but it lacks a major impractical advantage which mathematicians value highly and which, as we shall see, Lyapunov exponents do possess.

Chaos is defined in the long run. Uniform exponential growth of uncertainty is found only in the simplest chaotic systems. Indeed, uniform growth is rare amongst chaotic systems which usually display only **effective-exponential growth**, or equivalently *exponential-on-average* growth. The average is taken in the limit of an infinite number of iterations. The number we use to quantify this growth is call the *Lyapunov Exponent*. If the growth is a pure exponential, not just exponential-on-average, then we can quantify it as two raised to the power λ t, where t is time and λ is the Lyapunov exponent. The Lyapunov exponent has units of bits per iteration, and a positive exponent indicates the number of bits our uncertainty has grown *on average* after each iteration. A system has as many Lyapunov exponents as there are directions in its state space, which is the same as the number of components that make up the state. For convenience they are listed in decreasing order, and the first Lyapunov exponent, the largest one, is often called the *leading Lyapunov exponent*. In the sixties, the Russian mathematician Osceledec established that Lyapunov exponents existed for a wide variety of systems and proved that in many systems *almost all* initial conditions would share the same Lyapunov exponents. While Lyapunov exponents are defined by following the nonlinear trajectory of a system in state space, they only reflect the growth of uncertainty infinitesimally close to that nonlinear reference trajectory, and as long as our uncertainty is infinitesimal it can hardly damage our forecasts.

In as much as computing Lyapunov exponents requires averaging over infinite durations and restricts attention to infinitesimal uncertainties, adopting these exponents in the technical definition of mathematical chaos places this burden on identifying a system as chaotic. The advantage here is that these same properties make the Lyapunov exponent a robust reflection of the underlying dynamical system; we can take the state space and stretch it, fold it, twist it, and apply any smooth deformation, and the Lyapunov exponents do not change. Mathematicians prize that kind of consistency, and so Lyapunov exponents define whether or not a system has sensitive dependence. If the leading Lyapunov exponent is positive, then we have *exponential-on-average* growth of infinitesimal uncertainties, and a positive Lyapunov exponent is taken to be a necessary condition for chaos. Nevertheless, the same properties that give Lyapunov exponents their robustness make them rather difficult to measure in mathematical systems, and perhaps impossible to measure in physical dynamical systems. Ideally that should help us remain clear on the difference between mathematical maps and physical systems.

While there is no alternative with the mathematically appealing robustness of Lyapunov exponents, there are more relevant quantities for quantifying predictability. Knowing the average time it took a train to travel from Oxford to central London last week is more likely to provide insight into how long it will take today, than would dividing the distance between Oxford and London by the average speed of all trains which ever ran in England. Lyapunov exponents give us an average speed, while doubling times give us average times. By their very nature, Lyapunov exponents are far removed from any particular forecast.

Look at the menagerie of maps in Figure 8 (page 40): how would we calculate their Lyapunov exponents or doubling times? We wish to quantify the stretching (or shrinking) that goes on near a reference trajectory, but if our map is nonlinear then the amount of stretching will depend on how far we are from the reference trajectory. Requiring the uncertainty to remain infinitesimally close to the trajectory circumvents this potential difficulty. For one-dimensional systems we can then legitimately look at the slope of the map at each point. We are interested in how uncertainty magnifies with time. To combine magnifications we have to multiply the individual magnifications together. If my credit card bill doubles one day and then triples the next, the total increase is six times what I started with, not five. This means that to compute the average magnification per iteration we must take a ***geometric average***. Suppose the uncertainty increases by a factor of three in the first iteration, then by two, then by four, then by one third, and then by four: over all that is a factor of 32 over these five iterations: so on average the increase is by a factor of two per iteration, since the fifth root of 32 is two, that is: $2 \times 2 \times 2 \times 2 \times 2 = 32$. We are not interested in the arithmetic average: 32 divided by 5 is 6.4 and our uncertainty *never* grew that much on any one day. Also note that although the average growth is by a factor of two per day, the actual factors were 3, 2, 4, $\frac{1}{3}$, and 4: the growth was not uniform and on one day the uncertainty actually shrunk: if we can bet on the quality of our forecasts in a chaotic system and if we can bet different amounts on different days, then there may be times where we are *much* more confident in the future. Another myth bites the dust: chaos does not imply prediction is hopeless. In fact, if you can bet against someone who firmly believes that predicting chaos is uniformly hopeless, you are in a position to educate them.

The fact that some of the simplest cases (and most common examples) of chaos have constant slopes has lead to the overgeneralization that chaos is uniformly unpredictable. Looking back at the six chaotic systems in Figure 8 (page 40), we notice that in four of them (Shift Map, Tent Map, Quarter Map, and Tripling Tent Map), the magnitude of the slope is always the same. On the other hand, in the Logistic Map and the Moran-Ricker Map, the slope varies a great deal for different values of X. Since a slope with absolute value less than one indicates shrinking uncertainty, the Logistic Map shows strong growth of uncertainty at values of X near zero or near one, and shrinking of uncertainty for values of X near one-half! Likewise, the Moran-Ricker Map shows strong growth of uncertainty near zero and at values near one, where the magnitude of the slope is also large, but shrinking at intermediate and high values of X, where the slope is near zero.

How might we determine an average that extends into the infinite future? Like many mathematical difficulties, the easiest way to solve this one is to cheat. One reason that the Shift Map and the Tent Map are so popular in nonlinear dynamics is that while the trajectories are chaotic, the magnification of uncertainty is the same at each state. For the Shift Map, every infinitesimal uncertainty increases by a factor of two on each iteration. So the apparently intractable task of taking an average as time goes to infinity becomes trivial: if the uncertainty grows by a factor of two on every iteration then it grows by a factor of two on average, and the Shift Map has a Lyapunov exponent of one bit per iteration. Computing the Lyapunov exponent of the Tent Map is almost as easy: the magnification is either a factor of two or a factor of minus two, depending on which half of the 'tent' we are in. The minus sign does not effect the size of magnification: it merely indicates that the orientation has flipped from left to right, and we can safely ignore this. Again we have one bit per iteration. The same trick works for the Tripling Tent Map, but it has a larger slope of three, and a Lyapunov exponent of ~1.58 bits per iteration (the exact value is $\log_2(3)$). Why do we keep taking logarithms instead of just talking about 'magnifying factors' (Lyapunov numbers)? And why base 2 logarithms? This is a personal choice, usually justified by its connection to binary arithmetic, its use in computers, a preference for saying 'one bit per iteration' over saying 'about 0.693147 nats per iteration', and the fact that multiplying by two is relatively easy for humans.

The graph of the Full Logistic Map reveals a parabola, so the magnification at different states varies, and our trick of taking the average of a constant appears to fail. How might we take the limit into the infinite future? Our physicist would simply fire up a computer and compute finite-time Lyapunov exponents for many different states. Specifically, he would compute the geometric average magnification over two iterations for different values of X, and then the distribution corresponding to three iterations, then four iterations, ... And so on. If this distribution converges towards a single value, then he might be willing to count this as an estimate of the Lyapunov exponent, as long as the computer is not run so long as to be unreliable. As it turns out, this distribution converges faster than the Law of Large Numbers would suggest. Our physicist is happy with this estimated value, which turns out to be near one bit per iteration.

Our mathematician, of course, would not dream of making such an extrapolation. She sees no analogy between a finite number of digital computations, each of which is inexact, and an exact calculation extended into the infinite future. From her point of view, the value of the Lyapunov exponent at most values of α remains unknown, even today. But the Full Logistic Map is special, and demonstrates the second trick of mathematicians: substituting sin θ for X in the rule that defines the Full Logistic Map, and using some identities from trigonometry, she can show that the Full Logistic Map *is* the Shift Map. Since the Lyapunov exponents do not change under this kind of mathematical manipulation, she can prove that the Lyapunov exponent really is equal to one bit per iteration, and explain the violation of the Law of Large Numbers in a footnote.

Lyapunov exponents in higher dimensions

If the model state has more than one component, then uncertainty in one of its components can contribute to future uncertainty in other components. This brings in a whole new set of mathematical issues, since the order in which you multiply things together becomes important. We will initially avoid these complications by considering examples where the uncertainty in different components do not mix, but we must be careful not to forget that these are very special cases!

The state space of the **_Baker's Map_** has two components, x and y, as shown in Figure 21. It maps a two-dimensional square back into itself exactly with the rule:

> If x is less than one-half:
> Multiply x by 2 to get the new value of x and divide y by 2 to get the new y.
> Otherwise:
> Multiply x by 2 and subtract one to get the new value of x and divide y by 2 and add one half to get the new y.

In the Baker's Map, any uncertainty in the horizontal (x) component of our state will double on each iteration, while those in vertical (y) are cut in half. Since it is true on every step it is also true on average. The average uncertainty doubling time is one iteration and the Baker's Map has one Lyapunov exponent equal to one bit per iteration, and one exponent equal to minus one bit per iteration.

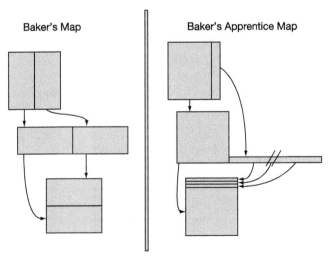

Baker's Map

Baker's Apprentice Map

21. Schematic showing how points in the square evolve forward under one iteration of (left) Baker's Map and (right) a Baker's Apprentice Map

The positive Lyapunov exponent corresponds to growing uncertainty, while the negative one corresponds to shrinking uncertainty. For every state, there is a direction associated with each of these exponents; in this very special case these directions are the same for all states and thus they never mix uncertainties in x with uncertainties in y. The Baker's Map itself was carefully crafted to avoid the difficulties caused by uncertainty in one component contributing to uncertainty in another component. In *almost all* two-dimensional maps, of course, such uncertainties do mix, so usually we cannot compute any positive Lyapunov exponents at all!

We can see why one might think predicting chaos is hopeless from the left panels of Figure 22, which show the evolution of a mouse-shaped ensemble over several iterations of the map. But remember that this map is a very special case: our hypothetical baker is very skilled in kneading, and can uniformly stretch the dough by a factor of two in the horizontal so that it shrinks by a factor of two in the vertical, before returning the lot back into the unit square. It is useful to contrast the Baker's Map with various members of the family of Baker's Apprentice Maps. Our hypothetical apprentices are each less uniform, stretching a small portion of the dough on the right side of the square a great deal, while hardly stretching the majority of the dough to the left at all, as shown in Figure 21. Luckily, all members of the Apprentice family are skilled enough not to mix the uncertainty in one component into another, so we can compute doubling times and Lyapunov exponents of any member.

As it turns out, every Apprentice Map has a leading Lyapunov exponent greater than the Baker's Map. So *if* we adopt the leading Lyapunov exponent as our measure of chaos, then the Apprentice Maps are each 'more chaotic' than the Baker's Map. This conclusion might cause some unease, when considered in light of Figure 22, which shows, side by side, the evolution of an ensemble of points under the Baker's Map and also under Apprentice number four. The

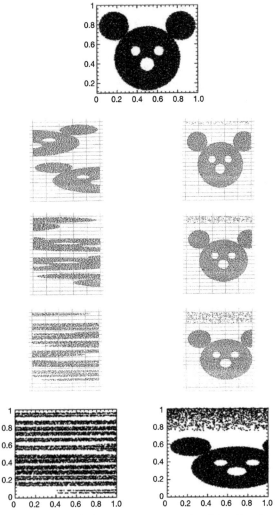

22. A mouse-like ensemble of initial states (top) and four frames,
showing in parallel the evolution of this ensemble under both the
Baker's Map (left) and the fourth Baker's Apprentice Map (right)

average doubling time of an Apprentice Map can be much greater than the Baker's Map, even though its Lyapunov exponent is also greater than that of the Baker's Map. This is true for an entire family of Apprentice Maps, and we can find an Apprentice Map with an average doubling time larger than any number one cares to name. Perhaps we should reconsider the connection between chaos and predictability?

Positive Lyapunov exponents with shrinking uncertainties

As long as our uncertainty is smaller than the smallest number we can think of, it can hardly pose any practical limit on our forecasts, and as soon as that uncertainty grows to be measurable, then its evolution need no longer be reflected by Lyapunov exponents in any way whatsoever. Even in the infinitesimal case, the Baker's Apprentice Maps show that Lyapunov exponents are misleading indicators of predictability, since the amount the uncertainty grows can vary with the state the system is in. And it gets better: in the classic system of Lorenz 1963 we can prove that there are regions of the state space in which all uncertainties *decrease* for a while. Given a choice as to when to bet on a forecast, betting when entering such a region will improve your odds of winning. Predicting chaotic systems is far from hopeless, betting against someone who naïvely believes it is hopeless might even prove profitable.

We end this discussion of Lyapunov exponents with one more word of caution. While a direction in which uncertainty neither grows nor shrinks implies a zero Lyapunov exponent the converse is not true: a Lyapunov exponent of zero does not imply a direction of no growth! Remember the discussion of the exponential that accompanied Fibonacci's rabbits: even growth as fast as the square of time is slower than exponential and will result in a zero Lyapunov exponent. This is one reason why mathematicians are so pedantic about really taking limits all the way out to the infinite future: if we consider a long but finite period of time, then *any* magnification at all would suggest a positive Lyapunov exponent – exponential, linear or even slower than linear growth will yield a magnification greater than one over any finite period, and the logarithm of any number greater than one will be positive. Computing the statistics of chaos will prove tricky.

Understanding the dynamics of relevant uncertainties

As we noted above, an infinitesimal uncertainty cannot cause us much difficulty in forecasting; once it becomes measurable, the details of its exact size and where the state is in the state space come into play. To date, mathematicians have found no elegant method for tracking these small but noticeable uncertainties, which are, of course, most relevant to real-world forecasting. The best we can do is to take a sample of initial states, called an ensemble, make this ensemble consistent both with the dynamics of our model and the noise in our observations, and then see how the ensemble disperses in the future. For our 21st-century demon that is enough: given her perfect model of the system and of the noise, her noisy observations of previous states reaching into the distant past, and her access to infinite computer power, her ensemble will accurately reflect the probability of future events. If a quarter of her ensemble members indicate rain tomorrow, then there really is a 25% chance of rain tomorrow, given the noisy observations available to her. Decreasing the noise increases her ability to determine what is more likely to happen. Chaos is no real barrier to her. She is uncertain of the present, but can accurately map that uncertainty into the future: who could ask for anything more? Our models, however, are not perfect and our computational resources are limited: in Chapter 9 we contrast the inadequacy with which we must deal with the uncertainty which she can accommodate.

The nonlinear zoo contains more than mere chaos. It need not be the case that the smaller the uncertainty, the more tame its behaviour. There are worse things than chaos: it could be the case that the smaller the uncertainty, the faster it grows, leading to an explosion of infinitesimal uncertainties to finite proportions after only a finite period. This is not as outlandish as it might sound: it remains an open question whether or not the basic equations of fluid dynamics display this worse-than-chaos behaviour – one of those few mathematical questions with a one million dollar reward attached to it!

Chapter 7

Real numbers, real observations, and computers

> The mathematician very carefully defines irrational numbers. The physicist never meets any such numbers ... The mathematician shudders at uncertainty and tries to ignore experimental errors.
>
> Leon Brillouin (1964)

In this chapter we examine the relation between the numbers in our mathematical models, the numbers we observe when taking measurements in the world, and the numbers used inside a digital computer. The study of chaos has helped to clarify the importance of distinguishing these three sorts of number. What do we mean by different kinds of number?

Whole numbers are integers; measurements of things like 'the number of rabbits in my garden' come naturally as integers, and computers can do perfect mathematics with integers as long as they do not get too big. But what about things like 'the length of this table', or 'the temperature at Heathrow Airport'? It seems these need not be integers, and it is natural to think of them as being represented by real numbers, numbers which can have an infinitely long string of digits to the right of the decimal point or bits to the right of the binary point. The debate over whether or not these real numbers exist in the real world dates back into antiquity. One thing that is clear is that when we 'take data' we only 'keep' integer values. If we measure 'the length of this table' and write it down as 1.370

the measurement does not appear to be an integer at first sight, but we can transform it into an integer by multiplying by 1000; anytime we are only able to measure a quantity like length or temperature to finite precision – which is always the case in practice – our measurement can be represented using an integer. And in fact our measurements are almost always recorded in this way today, since we tend to record and manipulate them using digital computers, which *always* store numbers as integers. This suggests something of a disconnect between our physical notion of length and our measurements of length, and there is a similar break between our mathematical models, which consider real numbers, and their computerized counterparts, which only allow integers.

Of course a real physicist would never say that the length of the table was 1.370; she would say something like the length was 1.370 ± 0.005, with the aim of quantifying her uncertainty due to noise. Implicit in this is a model of the noise. Random numbers from the bell-shaped curve is without doubt the most common noise model. One learns to include things like '± 0.005' in order to pass science classes in school; it is usually seen as an annoyance but what does it really mean? What is it that our measurements are measuring? Is there a precise number that corresponds to the True length of the table or the True temperature at the airport, but just obscured by noise and truncated when we record it? Or is it a fiction, and the belief that there should be some precise number just a creation of our science? The study of chaos has clarified the role of uncertainty and noise in evaluating our theories by suggesting new ways to see if such True values might exist. For the moment we will assume the Truth is out there and that we just cannot see it clearly.

Nothing really matters

So what is an observation exactly? Remember our first time series, which consisted of monthly numbers of rabbits in Fibonacci's mythical garden. In that case, we knew the total number of rabbits in the garden. But in most studies of population dynamics we do not have such complete information. Suppose for instance that we are studying a population of voles in Finland. We put out traps, check them each day, release the captives, and keep a daily time series of the number of voles captured. This number is somehow related to the number of voles per square kilometre in Finland, but how exactly? Suppose we observe zero voles in our trap today. What does this 'zero' mean? That there are no voles in this forest? That there are no voles in Scandinavia? That voles are extinct? Zero in our trap could mean any or none of these things and thus illustrates two distinct kinds of uncertainty we must cope with when relating our measurements to our models. The first is simple observational noise: an example would be to miscount the number of voles in the trap, or to find the trap full, leaving open the possibility that more voles might have been counted on that day if a larger trap had been used. The second is called *representation error*: our models consider the population density per square kilometre, but we are measuring the number of voles in a trap, so our measurement does not represent the variable our models use. Is this a shortcoming of the model or the measurement?

If we put the wrong number into our model we can expect to get the wrong number out: garbage in, garbage out. But it seems that our models are asking for one *kind* of number, while our observations are offering a noisy version of another kind of number. In the case of weather forecasting where our target variables – temperature, pressure, humidity – are thought to be real numbers, we cannot expect our observations to reflect the true values exactly. This suggests that we might look for models with dynamics which are *consistent* with our observations, rather than taking our observations and our model states to be more-or-less the same thing and trying to measure the distance between some future state of our model and the corresponding target observation. The goal of forecasting linear systems is to minimize this distance: the forecast error. When forecasting nonlinear systems it becomes important to distinguish the various things bound up in this quantity, including uncertainties in observation, truncation in measurement, and the difference between our mathematical models, our computer simulations of them, and whatever it was that actually generated the data. We first consider what happens when we try to put dynamics into a digital computer.

Computers and chaos

Recall that our three requirements for mathematical chaos were determinism, sensitive dependence, and recurrence. Computer models are deterministic to a fault. Sensitive dependence reflects the dynamics of infinitesimals, but on any given digital computer there is a limit to how close two numbers can be, beyond which the computer sees no difference at all and will treat them as if they were the same number. No infinitesimals, no mathematical chaos. A second reason that computers cannot display chaos arises from the fact there is only a finite amount of memory in any digital computer: each computer has a limited number of bits and thus only a limited number of different internal states, so eventually the computer must return to a state it has already been in, after which, being deterministic, the computer will simply run in circles, repeating its previous behaviour over and over forever. This fate cannot be avoided, unless some human or other external force interferes with the natural dynamic of the digital computer itself. A simple card trick illustrates the point nicely.

What does this imply for computer simulations of the Logistic Map? In the mathematical version of the map, the time series from iterating almost any X between zero and one will never contain the same value of X twice, no matter how many iterations we consider. As the number of iterations increases, the smallest value of X observed so far will slowly get closer and closer to zero, never actually reaching zero. For the computer simulation of the Logistic Map there are only about 2^{60} (about a million million million) different values of X between zero and one, so the time series from the computer must eventually include two values of X which are exactly the same, becoming stuck in an endless loop. After this

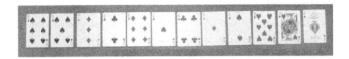

23. Two ways to deal the computers-can't-do-chaos card trick, if the deck of cards is large enough a time will come when everyone will find themselves on the same card even when they are placed on a line as in the upper panel

happens, the smallest value of X will never decrease again, and any computation along this loop, whether it be the average value of X or the Lyapunov exponent of the map, will reflect the characteristics of the particular loop, not the mathematical map. The computer trajectory has become *digitally periodic*, regardless of what the mathematical system would have done. And so it is for all digital computers. Computers cannot do chaos.

There may be more than one digitally periodic loop: shuffle a deck of cards and place some of them in a large circle so that the first card

Card tricks and computer programs

Ask a friend to pick a secret number between, say, 1 and 8, and then deal out a deck of cards as shown in Figure 23. Counting a face card as a ten and an ace as a one, ask your friend to count out her secret number and take the number of the card she lands on as her new number. If her number was one, she would land on the six of spades, and with the new number six, she would move forward to the four of clubs; if her original number was three, she would have hit the three of diamonds, then the ace of hearts, and so on. Try it yourself using Figure 23 and stop when you hit the jack of hearts. How did I know you would hit the jack of hearts? For the same reason that computers cannot display chaos. Everyone hits the jack of hearts.

What does this have to do with computers? A digital computer is a finite state machine: there are only a limited number of bits inside it which define its current state. Encoded in the current state of the machine is the rule that determines which state comes next. In the card game there were ten possible values at each location. If players on two different cards move forward to land on the same card, they remain identical from then on; unless one takes great care, nearby states in a computer will collapse in the same manner. A modern computer has many more options, but only a finite number, so eventually it will hit a configuration (an internal state) it has hit before, and after that happens it will cycle in the same loop forever. The card trick works in an analogous way: everyone starts off with their own initial number, updating and moving forward. But once two of these paths converge onto the same card, they stay together forever. For the particular cards on the table, everyone will hit the jack of hearts; no one will hit the ace of spades unless they start there. To see this, try starting with each value. If you pick one, you land on the six, then the four, then the jack; while picking two hits the five, the four, and the jack; picking three lands on the three, the ace, the four, and the jack; picking four, the two, the ace, the four, and the jack; picking five, the six and the jack; picking six, the ace, the four, the jack; picking seven, the four and the jack; picking 8, the ace, the two, and the jack. All values lead to the jack. Place the cards in a circle and we have a finite state machine where every starting point must lead to a periodic loop, but there may be more than one loop.

By projecting the cards on a screen, you can use this demonstration with a large audience. Take a number yourself, and deal out cards until you think everyone has converged. Then ask people to raise their hands if they are on, in this case, the jack of hearts. There is a wonderful look of surprise on the faces of the audience when they realize that they are all on the same card. They will converge faster if you restrict the deck to cards with small values. If you are willing to stack the deck to get more rapid convergence, what order would you place the cards?

follows the last card dealt. Determining which loop each card ends up in yields a list of all the loops. Which is larger: the number of cards that are actually on loops or those on transients? Shuffle the cards and repeat the experiment to see how the number of loops and their lengths change with the number of cards dealt. In the same way, artificially changing the number of bits a computer uses for each value of X turns it into a mathematical microscope for examining the digitally fine structure of the map, using the computer dynamics to examine the length scales where there would be far too many boxes to count them all.

Shadows of reality

> Reality is that which, when you stop believing in it, doesn't go away.
> P. K. Dick

Our philosopher and our physicist find these results disturbing. If our computers cannot reflect our mathematical models, how might we decide if our mathematical models reflect reality? If our computers cannot realize a mathematical system as simple as the Logistic Map, how can we evaluate the theory behind our much more complicated weather and climate models? Or contrast our mathematical models with reality? The issue of model inadequacy is deeper than that of uncertainty in the initial condition.

One test of model inadequacy is to take the observations we already have and ask if our model can generate a time series that stays close to these observations. If the model were perfect there would be at least one initial state that shadowed any length of observations we might take, where by **shadowing** we mean that the difference(s) between the model time series and the observed time series is consistent with our model for the noise. This gives our model for the noise a much higher status than it has ever had in the past. Can we still expect shadows when our models are not perfect? No, not in the long run, if our model is chaotic: we can prove that no shadowing trajectory exists. Noise will not go away, even when we stop believing in it. In imperfect chaotic models, we cannot get the noise to allow a coherent account of the difference between our models and the observations. Model error and observational noise are inextricably mixed together. And if observations, model states, and real numbers really are different kinds of number – like apples and orangutans – what did we think we were doing when we tried to subtract one from another? To pursue that question, we must first learn more about the statistics of chaos.

Chapter 8

Sorry, wrong number: statistics and chaos

I have no data yet, and it is a capital mistake to theorise before one has data.

(Holmes to Watson in *A Scandal in Bohemia*, A. C. Doyle)

Chaos poses new challenges to statistical estimation, but these need to be seen in the context of the challenges statisticians have been dealing with for centuries. When analysing time series from our models themselves, there is much to be gleaned from statistical insight and basic rules of statistical good practice. But our physicist faces an 'apples and oranges' problem when contrasting chaotic models with observations of the real world, and this casts the role of statistics in a less familiar context. The study of chaotic systems has clarified just how murky the situation is. There is even disagreement as to how to estimate the current state of a system given from noisy observations, which threatens to stop us from making a forecast before we even get started. Progress here would yield fruit on issues as disparate as our ability to foresee tomorrow's weather and our ability to influence climate change 50 years from now.

The statistics of limits and the limits of statistics

Consider estimating some particular statistic, say the average height of all human beings. There may be some disagreement over the definition of the population of 'all human beings' (those alive on 1 January 2000? those alive today? all those who have ever lived? . . .), but that need not distract us yet. Given the height of every member of this population a well-defined value exists, we just do not know what its value is. The average height taken over a sample of human beings is called the sample-average. All statisticians will agree on this value, even if they disagree about the relationship of this number to the desired average over the entire population. (Well, almost all statisticians will agree.) The same cannot be said for sample-Lyapunov exponents. It is not clear that sample-exponents of chaos can be uniquely defined in any sensible way.

There are several reasons for this. First, computing the statistics of chaos, like fractal dimensions and Lyapunov exponents, requires taking limits to vanishingly small lengths and over infinitely long durations. These limits can never be taken based on observations. Second, the study of chaos has provided new ways of making models from data without specifying exactly how to build them. The fact that different statisticians with the same data set may arrive at rather different **sample-statistics** makes the statistics of chaos rather different from the sample-mean.

Chaos changes what counts as 'good'

Many models contain 'free' parameters, meaning parameters which, unlike the speed of light or the freezing point of water, we do not already know with good accuracy. What then is the best value to give the parameter in our model? And if the purpose of the model is to make forecasts, why would we use a value from the lab or from some fundamental theory, if some other parameter value provided better forecasts? Modelling chaotic systems has even forced us to re-evaluate, arguably to redefine, 'better'.

In the weak version of the Perfect Model Scenario, our model has the same mathematical structure as the system which generated the data, but we do not know the True parameter values. Say we know that the data was generated by the Logistic Map, without knowing the value of α. In this case, there is a pretty well-defined 'best': the parameter value that generated the data. Given a perfect noise model for the observational uncertainty, how do we extract the *best* parameter values for use tomorrow given noisy observations from the past?

If the model is linear, then several centuries of experience and theory suggests the best parameters are those whose predictions fall closest to their targets. We have to be careful not to over-tune our model if we want to use it on new observations, but this issue is well known to our statistician. As long as the model is linear and the observational noise is from the bell-shaped distribution, then we have the intuitively appealing aim of minimizing the distance between the forecast and the target. Distance is defined in the usual least squares sense: based on adding up the squares of the differences in each component of the state. As the data set grows, the parameter values we estimate will get closer and closer to those that generated the data – assuming of course that our linear model really did generate the data. And if our model is nonlinear?

In the nonlinear case our centuries of intuition prove a distraction if not an impediment. The least squares approach can even steer us away from the correct parameter values. It is hard to understate the negative impact that failure to react to this simple fact has had on scientific modelling. There have been many warnings that things might go wrong, but given the lack of any clear and present danger – and their ease of use – such methods were regularly (mis)applied in nonlinear systems. Predicting chaos has made this danger clear: suppose we have noisy observations from the Logistic Map with (unknown to us) $\alpha = 4$, even with an infinite data set, the least squares approach yields a value for α which is too small. This is not a question of too little data or too little computer power: methods developed for linear systems give the wrong answer when applied to nonlinear questions. The mainstay of statistics simply does not hold when estimating the parameters of nonlinear models. This is a situation where ignoring the mathematical details and hoping for the best leads to disaster in practice: the mathematical justification for least squares assumes bell-shaped distributions for the uncertainty both on the initial state and on the forecasts. In linear models, a bell-shaped distribution for the uncertainty in the initial condition results in a bell-shaped distribution for the uncertainty in the forecasts. In nonlinear models this is not the case.

This effect is almost as important as it is neglected. Even today, we lack a coherent, deployable rule for parameter estimation in nonlinear models. It was the study of chaos that made this fact painfully obvious. Recently Kevin Judd, an applied mathematician at the University of Western Australia, has argued that not only the principle of least squares but the idea of maximum likelihood given the observations is also an unreliable guide in nonlinear systems. That does not imply that the problem is unsolvable: our 21st-century demon can estimate α very accurately, but she will not be using least squares. She will be working with shadows. Modern statistics is rising to the challenge of nonlinear estimation, at least in cases where the mathematical structure of our models is correct.

Lies, damn lies, and dimension estimates

A young student once had the intention,
to quantify fractal dimension.
But data points are not free,
and, needing 42-to-the-D,
she settled for visual inspection.

(after James Thieler)

While Mark Twain would probably have liked fractals, he would have without doubt hated dimension estimates. In 1983, Peter Grassberger and Itamar Procaccia published a paper entitled 'Measuring the Strangeness of Strange Attractors', which has now been cited by thousands of other scientific papers. Most papers gather only a handful of citations. It would be interesting to use these citations and examine how ideas from the study of chaos spread between disciplines, from physics and applied mathematics through every scientific genre.

The paper provides an engagingly simple procedure for estimating, from a time series, the number of components the state of a good model for a chaotic system would require. The procedure came complete with many well-signposted pitfalls. Nevertheless, many if not most applications to real data probably lie in one or the other of those pits. The mathematical robustness of the dimension is what makes capturing it such a prize: you can take an object and stretch it, fold it, roll it up in a ball, even slice it into a myriad of pieces and stick the pieces back together any old way, and you will not alter its dimension. It is this resilience that effectively requires huge data sets to have a fighting chance at meaningful results. Regrettably, the procedure tended towards false positives, and finding chaos by measuring a low dimension was fashionable. An unfortunate combination. Interest in identifying low-dimensional dynamics and chaos was triggered by a mathematical theorem, which suggested one might be able to predict chaos without even knowing what the equations were.

Takens' Theorem and embedology

Time series analysis was re-landscaped in the eighties as ideas from physicists in California led by Packard and Farmer were given a mathematical foundation by the Dutch mathematician Takens; with that basis new methods to analyse and forecast from a time series appeared apace. Takens' Theorem tells us that if we take observations of a deterministic system which evolves in a state space of dimension d, then under some very loose constraints there will be a nearly identical dynamical model in the delay space defined by *almost every* single measurement function (observation). Suppose the state of the original system has three components a, b, and c, the theorem says that one can build a model of the entire system from a time series of observations of any one of these three components; this is illustrated with real observations in Figure 24; taking just one measurement, say of a, and making a vector whose components are values of a in the present and in the past, results in a *delay-reconstruction* state space in which a model equivalent to the original system can be found. When this works, it is called a delay *embedding*. The 'almost every' restrictions are required to avoid picking a particularly bad period of time between the observations. By analogy: if you observed the weather only at noon, then you would have no inkling of what happened at night.

Takens' Theorem recasts the prediction problem from one of extrapolation in time to one of interpolation in state space. The traditional statistician who is at the end of his data stream and trying to forecast into an unknown future, while Takens' Theorem places our physicist in a delay-embedding state space trying to interpolate between previous observations. These insights impact more than just data-based models; complicated high-dimensional simulation models evolving on a lower-dimensional attractor might also be modelled by much lower-dimensional, data-based models. In principle, we could integrate the equations in this lower-dimensional space also, but in practice we set up our models as physical simulations in high-dimensional spaces; we can sometimes prove that lower-dimensional dynamics emerge, but we have no clue how to set up the equations in the relevant lower-dimensional spaces.

Comparing Figure 24 with Figure 14 makes it clear that the observations of the circuit 'look like' the Moore-Spiegel attractor, but how deep is this similarity, really? Every physical system is different. Often when we have little data and less understanding,

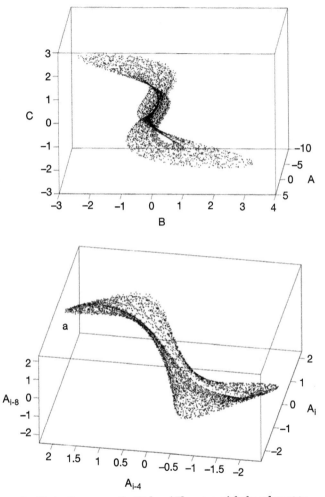

24. An illustration suggesting Takens' Theorem might be relevant to data from Machete's electric circuit carefully designed to produce time series which resemble those of the Moore-Spiegel System. Delay reconstruction of one measurement in the lower panel bears some resemblance to the distribution in the upper panel, which plots the values of three different simultaneous measurements. Contrast these with the lower panel of Figure 14 on page 67

then statistical models provide a valuable starting point for forecasting. As we learn more, and gather more data, simulation models often show behaviour 'similar' to the time series of observations, and as our models get more complicated this similarity often becomes more quantitative. On the rare occasions like this circuit when we have a vast duration of observations, it seems our data-based models – including those suggested by Takens' Theorem – often provide the best quantitative match. It is almost as if our simulation models are modelling some perfect circuit, or planet, while our data-based models more closely reflect the circuit on the table. In each case, we have only similarity; whether we use statistical models, simulation models, or delay-reconstruction models, the sense in which the physical system is described by any model equations is unclear. This is repeatedly seen in physical systems for which our best models are chaotic; we would like to make them empirically adequate, but are not always sure how to improve them. And with systems like the Earth's atmosphere, we cannot wait to take the required duration of observation. The study of chaos suggests a synthesis of these three approaches to modelling, but none has yet been achieved.

There are several common misinterpretations of Takens' Theorem. One is that if you have a number of simultaneous observations you *should* use only one of them; Takens allows us to use them all! A second is to forget that Takens' only tells us that *if* we have low-dimensional deterministic dynamics *then* many of its properties are preserved in a delay-reconstruction. We must be careful not to reverse the if-then argument and assume that seeing certain properties in a delay-reconstruction necessarily implies chaos, since we rarely if ever know the True mathematical structure of the system we are observing.

Takens' Theorem tells us that *almost any* measurement will work. This is a case where the 'almost any' in our mathematician's function space corresponds to 'not a single one' in the laboratories of the real world. Truncation to a finite number of bits violates an assumption of the theorem. There is also the issue of observational noise in our measurements. To some extent these are merely technical complaints; a delay reconstruction model may still exist and our statistician and physicist can rise to the challenge of approximating it given realistic constraints on the data stream. Another problem is more difficult to overcome: the duration of our observations needs to exceed the typical recurrence time. It may well be that the required duration is not only longer than our current data set, it may be longer than the lifetime of the system itself. This is a fundamental constraint with philosophical implications. How long would it take before we would expect to see two days with weather observations so similar we could not tell them apart? That is, two days for which the difference between the corresponding states of the Earth's atmosphere was within the observational uncertainty? About 10^{30} years. This can hardly be considered a technical constraint: on that time scale the Sun will have expanded into a red giant and vaporized the Earth, and the Universe may even have collapsed in the Big Crunch. We will leave our philosopher to ponder the implications held by a theorem that requires that the duration of the observations exceed the lifetime of the system.

In other systems, like a series of games of roulette, the time between observations of similar states may be much less. The search for dimensions from data streams is slowly being replaced by attempts to build models from data streams. It has been conjectured that it almost always takes less data to build a good model than it does to obtain an accurate dimension estimate. This is another indication that it may prove more profitable to pay attention to the dynamics rather than estimate statistics. In any event, the excitement of constructing these new data-based models brought many physicists into what had been largely the preserve of the statistician. A quarter of a century down the line, one major impact of Takens' Theorem was to meld the statisticians' approach to modelling dynamical systems with that of the physicists. Things are still evolving and a true synthesis of these two may yet emerge.

Surrogate data

The difficulty of getting to grips with statistical estimation in nonlinear systems has stimulated new statistical tests of significance using '*surrogate data*'. Scientists use surrogate data in a systematic attempt to break their favourite theories and nullify their most cherished results. While not every test that fails to kill a conclusion makes it stronger, learning the limitations of a result is always a good thing.

Surrogate data tests aim to generate time series which look like the observed data but come from a known dynamical system. The key is that this system is known *not* to have the property one is hoping to detect: can we root out results that look promising but in fact are not (called false positives) by applying the same analysis to the observed data and then to many surrogate data sets. We know at the start that the surrogate data can show only false positives, so if the observed data set is not easily distinguished from the surrogates, then the analysis holds few practical implications. What does that mean in practice? Well suppose we are hoping to 'detect chaos' and our estimated Lyapunov exponent turns out to be 0.5: is that value significantly greater than zero? If so then we have evidence for one of the conditions for chaos.

Of course, 0.5 is greater than zero. The question we want to answer is: are random fluctuations in an estimated exponent likely to be as big as 0.5 in a system (i) which produced similar-looking time series, and (ii) whose true exponent really was not greater than zero? We can generate a surrogate time series, and estimate the exponent from this surrogate series. In fact, we can generate 1,000 different surrogate series, and get 1,000 different exponents. We might then take comfort in our result if almost all of 1,000 estimates from the surrogate series are much less than the value of 0.5, but if the analysis of surrogate data often yields exponents greater than 0.5, then it is hard to argue that the analysis of the real data provided evidence for a Lyapunov exponent greater than zero.

Applied statistics

In a pinch, of course, one can drive a screw with a hammer. Statistical tools designed for the analysis of chaotic systems can provide a new and useful way of looking at observations from systems that are not chaotic. Just because the data do not come from a chaotic system does not mean that such a statistical analysis does not contain valuable information. The analysis of many time series, especially in the medical, ecological, and social sciences, may fall into this category and can provide useful information, information not available from traditional statistical analysis. Statistical good practice protects against being mislead by wishful thinking, and the insight obtained can prove of value in application, regardless of whether or not it establishes the chaotic credentials of the data stream.

Data Assimilation is the name given to translating a collection of noisy observations into an ensemble of initial model-states. Within PMS there is a True state that we can approximate, and given the noise model there is a perfect ensemble which, though available only to our 21st-century demon, we can still aim to approximate. But in all real forecasting tasks, we are trying to predict real physical systems using mathematical systems or computer simulations. The perfect model assumption is never justified and almost always false: What is the goal of data assimilation in this case? In this case, it is not simply that we get the 'wrong number' when estimating the state of our model that corresponds to reality, but that there is no 'right number' to identify. Model inadequacy appears to take even probability forecasts beyond our reach. Attempts to forecast chaotic systems with imperfect models are leading to new ways of exploring how to exploit the diversity of behaviour our imperfect models display. Progress requires we never blur the distinction between our mathematical models, our computer simulations and the real world that provides our observations. We turn to prediction in the next chapter.

Chapter 9
Predictability: does chaos constrain our forecasts?

> On two occasions I have been asked [by members of Parliament], 'Pray, Mr. Babbage, if you put into the machine wrong figures, will the right answers come out?' I am not able rightly to apprehend the kind of confusion of ideas that could provoke such a question.
>
> Charles Babbage

We are always putting the wrong numbers into our machines; the study of chaos has refocused interest in determining whether or not any 'right numbers' exist. Prediction allows us to examine the connection between our models and the real world in two somewhat different ways. We may test our model's ability to predict the behaviour of the system in the short term, as in weather forecasting. Alternatively, we may employ our models when deciding how to alter the system itself, here we are attempting to alter the future itself towards some desirable, or less undesirable, behaviour, as when using climate models for deciding policy.

Chaos poses no prediction problems for Laplace's demon: given exact initial conditions, a perfect model and the power to make exact calculations, it can trace a chaotic system forward in time as accurately as it can a periodic system. Our 21st-century demon has a perfect model and can make exact calculations, but is restricted to uncertain observations, even if they extend at regular intervals into the indefinite past. As it turns out, she cannot use these historical observations to identify the current state. She does, however, have access to the a complete representation of her uncertainty in the state given the observations that were made, some would call this an objective probability distribution for the state but we need not go there. These facts hold a number of implications: even with a perfect model of a deterministic system, she cannot do better than make probability forecasts. We cannot aspire to do better, and this implies that we will have to adopt probabilistic evaluation of our deterministic models. But all of these demons exist within the Perfect Model Scenario, we must abandon the mathematical fictions of perfect models and irrational numbers if we wish to offer honest forecasts of the real world. To fail to make it clear that we have done so would be to peddle snake oil.

Forecasting chaos

> And be these juggling fiends no more believ'd,
> That palter us in a double sense;
> That keep the word of promise to our ear,
> And break it to our hope.
>
> *Macbeth* (Act V)

Those who venture to predict have long been criticized even when their forecasts prove accurate, in a technical sense. Shakespeare's play *Macbeth* focuses on predictions which, while accurate in some technical sense, do not provide effective decision support. When Macbeth confronts the witches asking them what it is that they do, they reply 'a deed without a name'. A few hundred years later, Captain Fitzroy coined the term 'forecast'. There is always the possibility that a forecast be internally consistent from the modellers' perspective while actively misdirecting the forecast user's expectations. There lies the root of Macbeth's complaint against the witches: they repeatedly offer welcome tidings of what would seem to be a path to a prosperous future. Each forecast proves undeniably accurate, but there is little prosperity. Can modern forecasters who interpret uncertainty within their mathematical models as though it reflected real-world probabilities of future events hope to avoid the charge of speaking in a *double-sense*? Are they guilty of Macbeth's accusation in carefully wording their probability forecasts, knowing full well we will allow the excuse of chaos to distract us from entirely different goings on?

From accuracy to accountability

We can hardly blame our forecasters for failing to provide a clear picture of where we are going to end up at if we cannot give them a clear picture of where we are at. We can, however, expect our models to tell us how accurately we need to know the initial condition in order to ensure that the forecast errors stay below some target level. The question of whether or not we can reduce the noise to that level is, hopefully, independent of our model's ability to forecast given a sufficiently accurate initial state.

Ideally, a model will be able to shadow: there will be some initial state we can iterate so that the resulting time series remains close to the time series of observations. We have to wait until after we have the observations to see if a shadow exists, and 'close' must be defined by the properties of the observational noise. But if there is *no* initial state that shadows, then the model is fundamentally inadequate. Alternatively, if there is one shadowing trajectory there will be many. The collection of current states whose pre-histories have shadowed so far can be considered indistinguishable: if the True state is in there we cannot identify it. Nor can we know which of them will continue to shadow when iterated forward to form a forecast, but we could take some comfort from knowing the typical shadowing times of forecasts started from one of these indistinguishable states.

It is fairly easy to see that we are headed towards ensemble forecasts based upon candidates who have shadowed the observations up to the present. Realizing that even a perfect model couldn't yield a perfect forecast given an imperfect initial condition, in the 1960s, the philosopher Karl Popper defined an *accountable model* as one that could quantify a bound on how small the initial uncertainty must be in order to guarantee a specific desired limit on the forecast error. Determining this bound on the initial uncertainty is significantly more difficult for nonlinear systems than it is for linear systems, but we can generalize the notion of accountability and use it to evaluate whether or not our ensemble forecasts reasonably reflect probability distributions. Our ensembles will always have a finite number of members, and so any probability forecast we construct from them will suffer from this finite resolution: if we have 1,000 members then we might hope to see most events with a 1% chance of happening, but we know we are likely to miss events with only a 0.001% chance of happening. We will call an ensemble prediction system ***accountable*** if it tells us how big the ensemble has to be in order to capture events with a given probability. Accountability must be evaluated statistically over many forecasts, but this is something our statistician knows how to do quite well.

Our 21st-century demon can make accountable forecasts: she will not know the future but it will hold no surprises for her. There will be no unforeseeable events, and unusual events will happen with their expected frequency.

Model inadequacy

With her perfect model, our 21st-century demon can compute probabilities that are useful as such. Why can't we? There are statisticians who argue we can, including perhaps a reviewer of this book, who form one component of a wider group of statisticians who call themselves Bayesians. Most Bayesians quite reasonably insist on using the concepts of probability correctly, but there is a small but vocal cult among them that confuse the diversity seen in our models for uncertainty in the real world. Just as it is a mistake to use the concepts of probability incorrectly, it is an error to apply them where they do not belong. Let's consider an example derived from Galton's Board.

Look back at Figure 2 on page 9. You can buy modern incarnations of the image on the left on the internet, just Google 'quincunx'. The machine corresponding to the image on the right is more difficult to obtain. Modern statisticians have even questioned whether Galton actually built that one, although Galton describes experiments with the version, they have been called 'thought experiments' since even modern efforts to build a device to reproduce the expected theoretical results have found it 'exceedingly difficult to make one that will accomplish the task in a satisfactory manner'. It is not uncommon for a theorist to blame the apparatus when an experiment fails to match his theory. Perhaps this is merely an indication that our mathematical models are just different from the physical systems they aim to reflect? To clarify the differences between our models and reality, we will consider experiments on the Not A Galton (NAG) Board shown in Figure 25.

The NAG Board: an example of pandemonium

The NAG Board is 'Not A Galton Board'. It was originally constructed for a meeting to celebrate the 150th year of the Royal Meteorological Society, of which Galton was a member. The NAG Board has an array of nails distributed in a manner reminiscent of those in a Galton Board, but the nails are spaced further apart and imperfectly hammered. Note the small white pin at the top of the board, just to the left of half way. Rather than using a bucket of lead shot, golf balls are allowed through the NAG Board one at a time, each starting in exactly the same position, or as exactly as a golf ball can be placed under the white pin by hand. The golf balls do make a pleasant sound, but they do not make binary decisions at each nail; in fact, they occasional move horizontally past several nails before falling to the next level. Like the Galton Board and Roulette, the dynamics of the NAG Board are not recurrent: the dynamics of each ball is transient and so these systems do not display chaos. Spiegel suggested this behaviour be called *pandemonium*. Unlike the Galton Board, the distribution of golf balls at the bottom of the NAG Board does not reflect the bell-shaped distribution;

25. The Not A Galton Board, first displayed at a meeting held in St John's College, Cambridge, to celebrate the 150th year of the Royal Meteorological Society. Note the golf ball falling through the board does not make simple binary choices

nevertheless, we can use an ensemble of golf balls to gain a useful probabilistic estimate of where the next golf ball is likely to fall.

But reality is not a golf ball. Reality is a red rubber ball. And it is dropped only once. Laplace's demon would allow no discussion of what else might have happened: nothing else could have happened. The analogy here is to take the red rubber ball as the Earth's atmosphere and the golf balls as our model ensemble members. We can invest in as many members as we choose. But what does our distribution of golf balls tell us about the single passage of the red rubber ball? Surely the diversity of behaviour we observe between golf balls tells us something useful? If nothing else, it give us a lower bound on our uncertainty beyond which we know we cannot be confident; but it can never provide a bound in which we can be absolutely confident, even in probabilistic terms. By close analogy, examining the diversity of our models can be very useful, even if there is no probability forecast in sight.

The red ball is much like a golf ball: it has a diameter slightly larger but roughly the same as a golf ball and it has, somewhat more roughly, a similar elasticity. But the red ball which is reality can do things that a golf ball simply cannot do: some unexpected, some not; some relevant to our forecast, some not; some known, some not. In the NAG Board, the golf ball is a good model of reality, a useful model of reality; and an imperfect model of reality. How are we to interpret this distribution of golf balls? No one knows. Welcome to frontline statistical research. And it gets better. We could always interpret the distribution of golf balls as a probability forecast conditioned on the assumption that reality is a golf ball. Would it not be a double-sense to proffer probability forecasts one knew were conditioned on an imperfect model as if they reflected the likelihood of future events, regardless of what small print appeared under the forecast?

Our ensembles are not restricted to using only golf balls. We might obtain green rubber balls of a slightly smaller diameter and repeat the experiment. If we get a distribution of green balls similar to our distribution of golf balls, we might take courage – or better, take hope – that the inadequacies of our model might not play such a large role in the forecast we are interested in. Alternatively, our two models may share some systematic deficiency of which we are not aware . . . yet. But what if the distributions of golf balls and green balls are significantly different? Then we cannot sensibly rely on either. How might quantifying the diversity of our models with these multi-model ensembles allow us to construct a probabilistic forecast for the one passage of reality? When we look at seasonal weather forecasts, using the best models in the world, the distribution from each model tends to cluster together, each in a different way. How are we to provide decision support in this case, or a forecast? What should be our aim? Indeed, how exactly can we take aim at any goal given only empirically inadequate models? If we naïvely interpret the diversity of an ensemble of models as a probability, we will be repeatedly misled; we know at the start that our models are imperfect, so any discussion of 'subjective probability' is a red herring: we do not believe in (any of) our models in the first place!

The bottom line is rather obvious: if our models were perfect and we had the resources of Laplace's demon, we would know the future; while if our models were perfect and we had the resources of our 21st-century demon, then chaos would restrict us to probability forecasts, even if we knew the Laws of Nature were deterministic. In case the True Laws of Nature are stochastic, we can envision a statistician's demon, which will again offer accountable probability forecasts with or without exact knowledge of the current state of the universe. But is the belief in the existence of mathematically precise Laws of Nature, whether deterministic or stochastic, any less wishful thinking than the hope that we will come across any of our various demons offering forecasts in the woods?

In any event, it seems we do not currently know the relevant equations for simple physical systems, or for complicated ones. The study of chaos suggests that difficulty lies not with uncertainty in the number to 'put in' but the lack of an empirically adequate model to put anything into: chaos we might cope with, but it is model inadequacy, not chaos, that limits predictability. A model may undeniably be the best in the world, but that says nothing about whether or not it is empirically relevant, much less useful in practice, or even safe. Forecasters who couch predictions they expect to be fundamentally flawed with sleight-of-hand phrases such as 'assume the model is perfect' or 'best available information', may be technically speaking the truth, but if those models cannot shadow the past then it is not clear what 'uncertainty in the initial state' might mean. Those who blame chaos for the shortcomings of probability forecasts they devised under the assumption their models were perfect, models they knew to be inadequate, palter to us in a double-sense.

Chapter 10
Applied chaos: can we see through our models?

> All theorems are true,
> All models are wrong.
> All data are inaccurate.
> What are we to do?

Scientists often underestimate the debt they owe real-time forecasters who, day after day, stand up and present their vision of the future. Prominent among them are weather forecasters and economists, while professional gamblers risk more than their image when they go to work. As do futures traders. The study of chaos has initiated a rethink of modelling and clarified the restrictions on what we can see through our models. The implications differ, of course, for mathematical systems where we know there is a target to take aim at, and physical systems where what we aim for may well not exist.

Modelling from the ground up: data-based models

We will consider four types of data-based models. The simplest are *persistence models* which assume that things will stay as they are now. A simple dynamic variation on this theme are *advection models*, which assume the persistence of velocities: here, a storm moving to the east would be forecast to continue moving to the east at the same speed. Fitzroy and LeVerrier employed this approach in the 1800s, exploiting telegraph signals which could race ahead of an oncoming storm. The third are *analogue models*. Lorenz's classic 1963 paper ends with the sentence: 'In the case of the real atmosphere, if all other methods fail, we can wait for an analogue.' An analogue model requires a library of past observations from which a previous state similar to the current state is identified; the known evolution of this historical analogue provides the forecast. The quality of this method depends on how well we observe the state and whether or not our library contains sufficiently good analogues. When forecasting a recurrent system, obtaining a good analogue is just a question of whether or not the library's collection is large enough given our aims and the noise level. In practice, building the library may require more than just patience: how might we proceed if the expected time required to observe recurrence is longer than the lifetime of the system itself?

Traditional statistics has long exploited these three approaches within the context of forecasting from historical statistics. Takens' Theorem suggests that for chaotic systems we can do better than any of them. Suppose we wish to forecast what the state of the atmosphere will be tomorrow from a library. The situation is shown schematically in Figure 26. The analogue approach is to take the state in the library which is nearest to today's atmospheric state, and report whatever it did the next day as our forecast for tomorrow. Takens' Theorem suggests taking a collection of nearby analogues and interpolating between their outcomes to form our forecasts. These data-based *delay reconstruction models* can prove useful without being perfect: they need only outperform – or merely complement – the other options available to us. Analogue approaches remain popular in seasonal weather forecasting, while roulette suggests a data-based modelling success story.

It is easy to put money on a winner in roulette: all you have to do is bet one dollar on each number and you'll have a winner every time. You'll lose money, of course, since your winner will pay $36, while you'll have to bet on more than 36 numbers. 'Play them all'

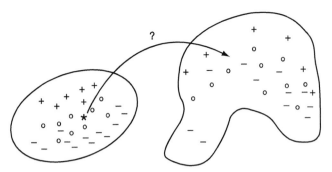

26. A schematic illustration of interpreting between analogues to make a forecast in a data-based state space. Knowing where the image of each nearby point falls, we can interpolate to form a forecast for the point marked '*'

strategies lose money on each and every game; casinos worked this out some time ago. Making a profit requires more than placing a winning bet every time: it requires a probabilistic forecast that is better than the house's odds. Luckily, that can be achieved short of the harsh requirements of empirical adequacy or mathematical accountability.

The fact that bets can be placed after the ball is in play makes roulette particularly interesting to physicists and the odd statistician. Suppose you record whenever the ball passes, say, the zero with the big toe on your left foot, and whenever zero passes a fixed point on the table with the big toe on your right foot; how often could a computer in the heel of your cowboy boot correctly predict which quarter of the roulette wheel the ball would land on? Predicting the correct quarter of the wheel half of the time would turn the odds in your favour: when you were right you'd win about four times the amount you lost, leaving a profit of three times your gamble, and you'd lose it all about half the time; so on average, you'd make about one and a half times the stake you put at risk. While the world will never know how many times people have tried this, we can put a lower bound of once: the story is nicely told by Thomas Bass in 'The Newtonian Casino'.

Simulation models

What if the most similar analogues did not provide a sufficiently detailed forecast? One alternative is to learn enough physics to build a model of the system from 'first principles'. Such models have proven seductively useful across the sciences, yet we must remember to come back from model-land and evaluate our forecasts against real observations. We may well have access to the best model in the world, but whether or not that model is of any value in making decisions is an independent question.

Figure 27 is a schematic reflecting the state space of a UK Met. Office Climate model. The state space of a numerical weather prediction (NWP) model falls along similar lines, but weather models are not run for as long as climate models, and so one often simplifies them by assuming things that change slowly, such as the oceans, sea ice or land use, are effectively constant. While the schematic makes models look more elaborate than the simple maps of previous chapters, once transfered onto a digital computer, the iteration of a weather model is not any more complex really, just more complicated. The atmosphere, along with the ocean, and the first few metres of the Earth's crust in some models, is effectively divided up into boxes; model variables – temperature, pressure, humidity, wind speed, and so on – are defined by one number in each box. In as much as it contains an entry for every variable in every grid-box, the model state can be rather large, some have over 10,000,000 components. Updating the state of the model is a straightforward if tedious process: one just applies the rule for each and every component, and iterates over and over again. This is what Richardson did by hand, taking years to forecast one day ahead. The fact that the calculations focus on components from 'nearby' cells gave Richardson the idea that a room full of computers arranged as shown in Figure 28 could in fact compute the weather faster than it happened. Writing in the 1920s, Richardson's computers were human beings. Today's multiprocessor digital supercomputers use more or less the same scheme. NWP models are among the most

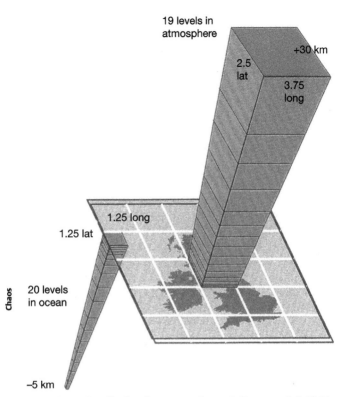

19 levels in atmosphere

+30 km

2.5 lat

3.75 long

1.25 long

1.25 lat

20 levels in ocean

−5 km

27. **A schematic reflecting the way weather and climate models divide both the atmosphere and the ocean into "grid points". Here each grid point in the atmosphere represents approximately a 250 km by 250 km square, which means that about six points account for the whole of Britain as shown above**

complicated computer codes ever written and often produce remarkably realistic-looking simulations. Like all models, however, they are imperfect representations of the real-world system they target, and the observations we use to initialize them are noisy. How are we to use such valuable simulations in managing our affairs? Can we at least get an idea of how much we should rely on today's forecast for next weekend?

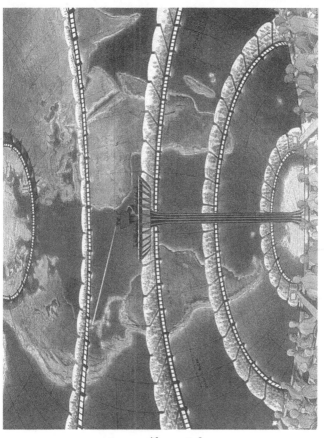

28. A realization of Richardson's dream, in which human computers work in massively parallel style to calculate the weather before it happens. Note the director in the central platform is shining a light on northern Florida, presumably indicating that those computers are slowing down the project (or perhaps the weather there is just particularly tricky to compute?)

Ensemble weather prediction systems

Latest EPS giving northern France an edge over Cornwall. Do you have a travel agent who can advise on ferry bookings?   Tim

email dated 5 August 1999

In 1992 operational weather forecasting centres on both sides of the Atlantic took a great step forward: they stopped trying to say exactly what the weather would be next weekend. For decades, they had run their computer simulations once a day. As computers grew faster, the models had grown more and more complicated, limited only by the need to get the forecast out well before the weather arrived. This 'best guess' mode of operation ended in 1992: instead of running the most complicated computer simulation once and then watching as reality did something different, a slightly less complex model was run a few dozen times. Each member of this ensemble was initialized at a slightly different state. The forecasters then watched the ensemble of simulations spread out from each other as they evolved in time towards next weekend, and used this information to quantify the *reliability* of the forecast for each day. This is an Ensemble Prediction System (EPS).

By making an ***ensemble forecast*** we can examine alternatives consistent with our current knowledge of the atmosphere and our models. This provides significant advantages for informed decision support. In 1928, Sir Arthur Eddington predicted a solar eclipse 'visible over Cornwall' for 11 August 1999. I wanted to see this eclipse. So did Tim Palmer, Head of the Probability Forecast Division at the European Centre for Medium-range Weather Forecasts (ECMWF) in Reading, England. As the eclipse approached, it seemed Cornwall might be overcast. The email from Tim quoted at the beginning of this section was sent six days before the eclipse: we examined the ensemble for the 11th, noting that the number of ensemble members suggested clear sky over France exceeded the corresponding number for Cornwall; the same thing happened on the 9th and we left England for France by ferry.

There we saw the eclipse, thanks to playing the odds suggested by the EPS, and to a last minute dash for better visibility made possible by Tim's driving skills on tiny French farm roads in a right-hand-drive car; not to mention his solar eclipse black-out glasses. The study of chaos in our model suggests that our uncertainty in the current state of the atmosphere makes it impossible to say for certain, even only a week in advance, where the eclipse will be visible and where it will be obscured by clouds. By running an ensemble forecast with the aim of tracking this uncertainty, the EPS provided effective decision support nevertheless: we saw the eclipse. We did not have to assume anything about the perfection of the model and there were no probability distributions in sight.

Since the EPS first became operational in 1992, no ensemble forecast was generated for the Burns' Day storm of January 1990. ECMWF has kindly generated a retrospective ensemble forecast using the data available two days before the Burns' Day storm struck. Figure 4 (on page 14) shows the storm as seen within a modern weather model – called the *analysis* – along with a two-day-ahead forecast using only data from before the time of the critical ship observations discussed in Chapter 1. Note that there is no storm in the forecast. Twelve other ensemble members also from two days before the storm are shown in Figure 29; some have storms, some not. The second ensemble member in the top row looks remarkably like the analysis; the member two rows below it has what looks like a super-storm while other members suggest a normal British winter's day. As the critical ship observations were made after this EPS forecast, this ensemble would have already provided an indication that a storm was likely, and significantly reduced the pressure on the intervention forecaster. At longer lead times, the ensemble from three days before Burns' Day has members with storms over Scotland, and there is even one member from the four-day-ahead ensemble forecast with a major storm in the vicinity. The ensemble provides early warning.

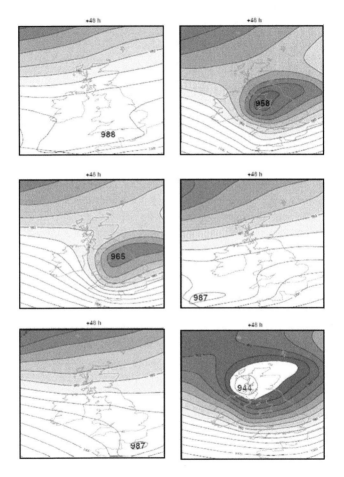

29. An ensemble of forecasts from the ECMWF weather model, two days in advance of the Burns' Day storm: some show storms, some do not. Unlike the single 'best guess' forecast shown in Figure 4 on page 14, here we have some forewarning of the storm

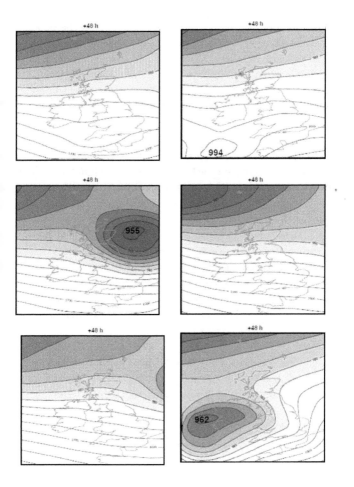

At all lead times, we must cope with the Burns effect: our collection of ECMWF weather 'golf balls' shows the diversity of our model's behaviour to aid us when we 'guess and fear', without actually quantifying the uncertainty in our real-world future. In fact, we could widen this diversity, if we have enough computer power and questioned the reliability of certain observations, we might run some ensemble members with those observations while omitting them in others. We will never see another situation quite like the Burns' Day storm of 1990. We might decide where to take future observations designed to maximize the chance of distinguishing which of our ensemble members were most realistic: those with a storm in the future or those without?

Rather than wasting too much energy trying to determine the 'best' model, we might learn that ensembles members from different models were of more value than one simulation of an extremely expensive super-model. But we should not forget the lessons of the NAG Board: our ensembles reveal the diversity of our models not the probability of future events. We can examine ensembles over initial conditions, parameter values, even mathematical model structures, but it seems only our 21st-century demon can make probability forecasts which are useful as such. Luckily, an EPS can inform and add value without providing probabilities that we would use as such for decision making.

Just after Christmas in 1999, another major storm swept across Europe. Called T1 in France and Lothar in Germany, this storm destroyed 3,000 trees in Versailles alone and set new record high insurance claims in Europe. Forty-two hours before the storm, ECMWF ran its usual 51-member EPS forecast. Fourteen members of the 51-member ensemble had storms. It is tempting to forget these are but as golf balls on a NAG Board, and interpret this as saying that there was about a 28% probability of a major storm. Even though that temptation should be resisted, we have here another EPS forecast with great utility. Running a more realistic, more complicated model once might have shown a storm, or might have shown no storm: why take the chance of not seeing the storm when an EPS might quantify that chance? Ensemble forecasting is clearly a sensible idea, but how exactly should we distribute limited resources between using a more expensive model and making a larger ensemble? This active research question remains open. In the meantime, the ECMWF EPS regularly provides a glimpse of alternative future scenarios seen through our models with significant added value.

How to communicate this information in the ensemble without showing the public dozens of weather maps also remains an open question. In New Zealand, where severe weather is rather common, the Meteorological Service regularly makes useful probabilistic statements on their website – statements like 'two chances in five'. This adds significant value to the description of a likely event. Of course, meteorologists often display a severe weather fetish, while energy companies are happy to exploit the significant economic value in extracting useful information from more mundane weather, every day. And those in other sectors with operational weather risk are beginning to follow suit.

Chaos and climate change

Climate is what you expect. Weather is what you get.
Robert Heinlein, Time Enough for Love (1974)

Climate modelling differs fundamentally from weather forecasting. Think of the weather in the first week in January a year from now. It will be mid-summer in Australia and mid-winter in the northern hemisphere. That alone gives us a good idea of the range of temperatures to expect: this collection of expectations is climate – ideally reflecting the relative probability of every conceivable weather pattern. If we believe in physical determinism, then the weather next January is already preordained; even so, our concept of the climate collection is relevant, as our current models are not able to distinguish that preordained future. The ideal ensemble weather forecast would trace the growth of any initial uncertainty in the state of the atmosphere until it became indistinguishable from the corresponding climate distribution. Given imperfect models, of course, this doesn't ever quite happen, as our ensemble of model simulations evolves towards the attractor of the model not that of the real world, if such a thing exists. Even with a perfect model, and ignoring the impacts of human free will noted by Eddington, accurate probability forecasts based on the current conditions of the Earth would be prevented by influences just now leaving the Sun, or those due to arrive from beyond the solar system, of which we cannot know today, even in principle.

Climate modelling also differs from weather forecasting in that it often contains a 'what if' component. Altering the amount of carbon dioxide (CO_2) and other greenhouse gases in the atmosphere is analogous to changing the parameter a in the Logistic Map; as we change parameter values, the attractor itself changes. In other words, while weather forecasters try to interpret the implications a distribution of golf balls holds for the single drop of a red rubber ball in the NAG Board of Figure 25 (page 128), climate modellers add the complication of asking what would happen if the nails were moved about.

Looking at just one run of a climate model carries the same dangers as looking at just one forecast for Burns' Day in 1990, although the repercussions of such naïve over-confidence would be much greater in the climate case. No computing centre in the world has the power to run large ensembles of climate models. Nevertheless, such experiments are made possible by harnessing the background processing power of PCs in homes spread about the globe (see *www.climateprediction.net*). Thousands of simulations have revealed that a surprisingly large range of diversity exists within one state-of-the-art climate model, suggesting that our uncertainty in the future of real-world climate is at least very large. These results contribute to improving current models. They fail to provide evidence that the current generation of climate models can realistically focus the questions of regional detail, which, when available, will be of great value in decision support. A frank appraisal of the limitations of today's climate models casts little doubt upon the wide consensus that significant warming has been seen in the data of the recent past.

How wide is the current diversity among our models? This depends, of course, on what model variables you examine. In terms of planet-wide average temperature, there is a consistent picture of warming; a goodly number of ensemble members show a great deal more warming than was previously considered. In terms of regional details, there are vast variations between ensemble members. It is hard to judge the utility of estimated precipitation for decision support, even for monthly rainfall over the whole of Europe. How might one distinguish what are merely the best currently available forecasts from forecasts that actually contain useful information for decision makers in the climate context?

In reality, carbon dioxide levels and other factors are constantly changing, weather and climate merge into a single realization of a one-off transient experiment. Weather forecasters often see themselves as trying to extract useful information from the ensemble before it spreads out across the 'weather attractor'; climate modellers must address difficult questions about how the structure of that attractor would change if, say, the amount of carbon dioxide in the atmosphere was doubled and then held constant. Lorenz was already doing active research here in the 1960s, warning that issues of structural stability and long transients complicate climate forecasts, and illustrating the effects in systems not much more complicated than the maps we defined in Chapter 3.

Given that our weather models are imperfect, their ensembles do not actually evolve towards realistic climate distributions. And given that the properties of the Earth's climate system are constantly changing, it makes little sense to talk about some constantly changing, unobservable 'realistic climate distribution' in the first place. Could any such thing exist outside of model-land? That said, coming to understand chaos and nonlinear dynamics has improved both the experimental design in and the practice of climate studies, allowing more insightful decision support for policy makers. Perhaps most importantly, it has clarified that difficult decisions will have to be made under uncertainty. Neither the fact that this uncertainty is not tightly constrained nor the fact that it can only be quantified with imperfect models, provides an excuse for inaction. All difficult policy decisions are made in the context of the Burns effect.

Chaos in commerce: new opportunities in Phynance

When a large number of people are playing a game with clear rules but unknown dynamics, it is hard to distinguish those who win with skill from those who win by chance. This is a fundamental problem in judging hedge-fund managers and improving weather models, since traditional scores can actually penalize skilful probabilistic play. The Prediction Company, or PredCo, was founded on the premise that there must be a better way to predict the economic markets than the linear statistical methods that dominated quantitative finance two decades ago. PredCo set out upon a different path blazed by Doyne Farmer and Norm Packard, along with some of the brightest young nonlinear dynamicists of the day, who gave up post-docs for stock options. If there was chaos in the markets, perhaps others were being be fooled without randomness? Sadly, confidentiality agreements still cloud even PredCo's early days, but the continued profitability of the company indicates that whatever it is doing, it is doing it well.

PredCo is one example of a general move towards Phynance, bringing well-trained mathematical physicists in to look at forecast problems in finance, traditionally the statistician's preserve. Is the stock market chaotic? Current evidence suggests our best models of the markets are fundamentally stochastic, so the answer is 'no'. But neither are they linear. To take one example, the study of chaos has contributed to fascinating developments at the interface of weather and economics: many markets are profoundly affected by weather, some are even affected by weather forecasts. Many analysts so fear that they might be fooled by randomness that they are religiously committed to fairly simple, purely stochastic models, and ignore the obvious fact that some ensemble weather forecasts contain useful information. For energy companies, information on the uncertainty of weather information is being used daily to avoid 'chasing the forecast': buying high, then selling low, then buying high the same cubic metre of natural gas yet again as the weather forecast for next Friday's temperature jumps down, then up, then down again, taking the expected electricity demand for next Friday along with it at each jump. That fact has put speculators in hot pursuit of methods to forecast the next forecast.

The study of chaos leads to efficiency beyond short-term profit; Phynance is making significant contributions to the improved distribution of perishable goods with weather-related demand, ship, train and truck transport, and demand forecasting in general. Better probabilistic forecasts of chaotic fluctuations in wind and rain significantly increase our ability to use renewable energy, reducing the need to keep fossil fuel generators running on 'standby', except on days of truly low predictability.

Retreating towards a simpler reality

Physical systems inspired the study of chaotic dynamical systems, and we now understand how our 21st-century incarnation of Laplace's demon could generate accountable probability forecasts of chaotic systems with her perfect model. Whether purely data-based or derived from today's 'Laws of Nature', the models we have to hand are imperfect. We must contend both with observational uncertainty and with model inadequacy. Interpreting an ensemble forecast of the real world as if it were a perfect model probability forecast of a mathematical system is to make the most naïve of forecasting blunders. Can we find a single real-world system in which chaos places the ultimate limit on our forecasts?

The Earth's atmosphere/ocean system is a tough forecasting nut to crack; physicists avoid a complete retreat to mathematical models by examining simpler physical systems on which to break their forecasting procedures and theories of predictability. We will track the course of this retreat from the Earth's atmosphere to examine the last ditch, and then examine what lies there in some detail. Lorenz noted the laboratory 'dish pan' experiments of Raymond Hide to support chaotic interpretations of his computer simulations in the early 1960s. Offspring of those experiments are still rotating in the Physics Department of Oxford University, where Peter Read provides the raw material for their data-based reconstructions. Thus far, probabilistic forecasts of these fluid systems remain very imperfect. Experimentalists around the globe have taken valuable data both from fluid systems and from mechanical systems motivated by the chaotic nature of the corresponding physical models. Real pendulums tend to heat up, changing the 'fixed' parameters of simulation models while leaving the regions of state space on which data-based models were trained. Even dice wear down, a bit, on each throw. Such is the nature of the real world.

Physical systems providing large volumes of data, low observational noise levels, and physically stationary conditions might prove more amenable to the tools of modern nonlinear data analysis. Ecosystems are right out. Fast, clean, and accurately instrumented lasers have proven rich sources, but we do not have accountable forecast models here or when studying the dynamics of more exotic fluids like helium. At the last ditch we find electronic circuits: arguably simple analogue computers. A manuscript reporting successful ensemble forecasts of these systems is likely to be rejected by professional referees for having taken too simple a system. So much greater the insight when we *fail* to generate accountable forecasts for these simplest of real-world systems.

Figure 30 shows ensemble forecasts of observed voltages in a circuit constructed to mimic the Moore-Spiegel system. Forecasts from two different models are shown. In each panel, the dark solid line shows the target observations generated by the circuit, while each light line is an ensemble member; the forecasts start at time equals zero; the ensemble was formed using only observations taken before that time. The top two panels show results from Model One, while the bottom two show results from Model Two. Look at the two panels on the left, which show simultaneous forecasts from each model. Every member of the Model One ensemble runs away from reality without warning just before time 100, as shown in the upper panel; the Model Two ensemble in the lower panel manages to spread out at about the correct time (or is it a bit early?), and the diversity of this ensemble looks useful all the way to the end of the forecast. In this case, we may not know which model was going to prove correct, but we can see where they began to strongly diverge from each other. On the panels to the right, both models fail at about the same time, in about the same way.

In each case, it appears that the forecasts provide insight into the likely future observations, but that the point in the future when this insight fails is not well reflected by either ensemble system. How can we best interpret this diversity in terms of a forecast?

Analysis of many forecasts from different initial conditions shows that, interpreted as probability forecasts, these ensembles are not accountable. This seems to be a general result when using arguably chaotic mathematical models to forecast real-world systems. I know of no exceptions. Luckily, utility does not require extracting useful probability estimates.

Odds: do we really have to take our models so seriously?

In academic mathematics, odds and probabilities are more or less identical. In the real world this is not the case. If we add up the

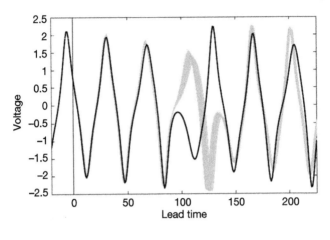

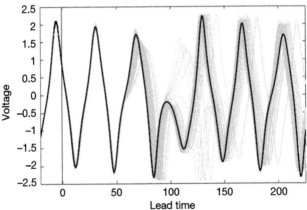

30. **Ensemble forecasts of the Machete's Moore-Spiegel Circuit. The dark line shows the observations; the light lines are the ensemble members; the forecast starts at time zero. The two panels on the left show ensemble forecasts for the same data but made by two different models; note that the ensemble in the lower panel manages to catch the circuit even when the model in the upper panel loses it near time 100. Forecasts from a second initial condition by these same two models is shown in the two panels on the right where the ensembles under both models fail at about the same time**

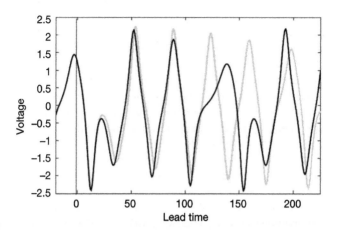

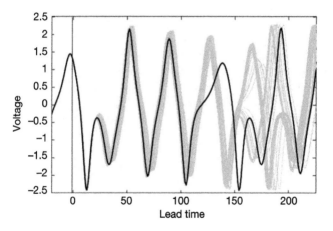

probability of every possible event, then the sum of the probabilities should be one. For any particular set of odds-on, we can then define the *implied probability* of an event from the odds on that event. If the sum of the implied probabilities is equal to one, then this set of odds are *probabilistic odds*. Outside mathematics lectures, probabilistic odds are rather hard to find in the real world. The related notion of 'fair odds', where the odds are fixed and one is given the option to take either side of a bet, suggests a similar sort of ivory tower 'wishful thinking'; implied probabilities from odds-against do not complement those from odds-on. The confusion at the heart of both conceptions comes largely from blurring the distinction between mathematical systems and the real-world systems they model. At the racetrack or in a casino, the implied probabilities sum to more than one. A European roulette wheel yields 37/36, while an American wheel yields 38/36. In a casino this excess ensures profit; scientifically, we might exploit this same excess to communicate information about model inadequacy.

Model inadequacy can steer us away from probability forecasts in a manner not dissimilar to the way in which uncertainty in the initial condition steers us away from the principle of least squares in nonlinear models. Theory for incorporating probability forecast systems into a decision support by maximizing expected utility – or some other reflection of the user's goal – is well developed. A 'probability forecast' which would not be used as such in this setting should perhaps not be called a probability forecast at all. A theory for incorporating forecast systems which provides odds rather than probabilities for decision support could, no doubt, be constructed. Judd has already provided several worked examples.

It appears that accepting the inadequacy of our own models, while being ignorant of the inadequacy of the models to which the competition has access, requires we aim for something short of fair odds. If an odds prediction system can cover its losses – breaking even when evaluated over all comers while covering its running costs – then we can say it generates *sustainable odds*. Sustainable odds then provide decision support which does not result (has not yet resulted) in catastrophe nor instilled the desire to invest more in improving those odds in order either to gain greater market share or to cover running expenses.

Ensembles over all the alternatives one can think of to sample might lead to sustainable odds, allowing the diversity within multi-model ensembles to estimate the impact of model inadequacy. The extent to which the sum of our implied probabilities exceeds one provides a manner to quantify model inadequacy. One wonders if, as we understand some real-world system better and better, we can expect the implied probabilities of our odds forecasts to ever sum to one for *any* physical system?

Moving to forecast systems which provide odds rather than probabilities releases our real-world decision support from unnatural constraints due to probabilities, which may be well-defined only in our mathematical systems. It is an awkward but inescapable fact that sustainable odds will depend both on the quality of your model and on that of the opposition. Decision making would be easy if accountable probability forecasts were on offer, but when model diversity cannot be translated into (a decision relevant) probability, we have no access to probability forecasts. Pursuing risk management as if we did for the sake of simplicity is foolhardy. And while odds might prove useful in hourly or daily decision making, what are we to do in the climate change scenario, where it appears we have only one high-impact event and no truly similar test cases to learn from?

We have reached the coal face of real-world scientific forecasting. The old seam of probability is growing thin and it is unclear exactly which direction we should dig in next. If chaotic dynamical systems have not provided us with a new shovel, they have at least given us a canary.

Chapter 11
Philosophy in chaos

You don't have to believe everything you compute.

Is there really anything new in chaos? There is an old joke about three baseball umpires discussing the facts of life within the game. The first umpire says 'I calls'em as I see'em.' The second umpire says 'I calls'em as they are.' Finally, the third says 'They ain't, until I calls'em.' The study of chaos tends to force us towards the philosophical position of the third referee.

Complications of chaos

Do the quantities we forecast exist only within the forecast models we construct? If so, then how might we contrast them with our observations? A forecast lies in the state space of our model and, while the corresponding observation is not in that state space, are these two 'subtractable'? This is a mathematical version of the 'apples and oranges' problem: are the model state and the observation similar enough that we can meaningfully subtract one from the other to define a distance, to then call a forecast error? Or are they not? And if not, then how might we proceed?

Evaluation of chaotic models has exposed a second fundamental complication that arises even in perfect nonlinear models with unknown parameter values: how do we determine the best values? If the model is linear, then we have several centuries of experience and theory which convincingly establish that the best values in practice are those that yield the closest agreement on the target data, where closest is defined in a least squares sense (smallest distance between the model and the target observations); likelihood is a useful thing to maximize. If our model is not linear, then our centuries of intuition often prove a distraction, if not an impediment to progress. Taking least squares is no longer optimal, and the very idea of 'accuracy' has to be rethought. This simple fact is as important as it is neglected. This problem is easily illustrated in the Logistic Map: given the correct mathematical formula and all the details of the noise model – random numbers with a bell-shaped distribution – using least squares to estimate α leads to systematic errors. This is not a question of too few data or insufficient computer power, it is the method that fails. We can compute the optimal least squares solution: its value for α is too small at all noise levels. This principled approach just does not apply to nonlinear models because the theorems behind the principle of least squares repeatedly assume bell-shaped distributions. The shape of these distributions is preserved by linear models, but *nonlinear models distort the bell-shape*, making least squares inappropriate. In practice, this 'wishful linear thinking' systematically underestimates the true parameter value at every noise level. Recent (mis)interpretations of climate models have floundered due to similarly wishful linear thinking. Our 21st-century demon will be able to estimate α very accurately, but she will not be using least squares to do so! (She will be looking for shadows.)

Philosophers have also wondered whether fractal intricacy might establish the existence of real numbers in nature, proving that irrational numbers exist even if we can only see a few of the leading bits. Strange attractors offer nothing to support such arguments that cannot be obtained from linear dynamical systems. On the other hand, chaos offers a new way to use both models and our observations to define variables in remarkable detail – if our models are good enough – via states along the shadow from an empirically adequate nonlinear model. If our model shadows the observations for an extended time, then all the shadowing states will fall into a very narrow range of values, providing a way to define values for observables like temperature to a precision beyond that at which our usual concept of temperature breaks down. We will never get to an irrational number, but an empirically adequate model could supply a definition of arbitrary accuracy, using the observations while placing the model into a role not unlike that of the third umpire. That said, the traditional connection between temperature and our measurements of it via a noise model, remains safe until useful shadowing trajectories are shown to exist.

Another philosophical quandary arises in terms of how to define the 'best' forecast in practice. Probabilistic forecasts provide a distribution as each forecast, while the target observation we verify against will always be a single event: when the forecast distribution differs from one forecast to the next, we have yet another 'apples and oranges' problem and can never evaluate even one of our forecast distributions as a distribution.

The success of our models tends to lull us towards the happy thought that mathematical laws govern the real-world systems of interest to us. Linear models formed a happy family. The wrong linear model can be close to the right linear model, and seen to be so, in a sense that does not apply to nonlinear models. It is not easy to see that an imperfect nonlinear model is 'close to' the right model given only observations: we can see that it allows long shadows, but if the two models have different attractors – and we know that the attractors of very similar mathematical models can be very different – then we do *not* know how to make ensembles that produce accountable probability forecasts. We must reconsider how our nonlinear models might approach Truth, in the case that Truth can be encapsulated in some 'right' model. We have no scientific reason to believe that such a perfect model exists. Our philosopher might turn from muddy issues raised on the quest for Truth and contemplate the implications of there being nothing more than collections of imperfect models. What advice might she offer our physicist? If new computer power allows the generation of ensembles over everything we can think of (initial conditions, parameter values, models, compilers, computer architecture, and so on), how do we interpret the distributions that come out scientifically? Or expose the folly of hiding from these issues behind a single simulation from a particularly complicated ultra-high-resolution model?

Lastly, note that when working with the wrong model, we may ask the wrong question. Who is who in la Tour's card game? The question assumes a model in which each player can be only a mathematician or a physicist or a statistician or a philosopher, and that there must be a representative of each discipline at the table. Perhaps this assumption is false. As real-world scientists, can each of our players take on every role?

The burden of proof: what is chaotic, really?

If we stay with mathematical standards of proof, then very few systems can be proven to be chaotic. The definition of mathematical chaos can only be applied to mathematical systems, so we cannot begin to prove a physical system is chaotic, or periodic for that matter. Nevertheless, it is useful to describe physical systems as periodic or chaotic as long as we do not confuse the mathematical models with the systems we use them to describe. When we have the model in hand, we can see whether it is deterministic or stochastic, but even after knowing it to be deterministic, proving it to be chaotic is non-trivial. Calculating Lyapunov exponents is a difficult task, and there are very few systems for which we can do this analytically. It took almost 40 years to establish a mathematical proof that the dynamics of the 1963 Lorenz System were chaotic, so the question regarding more complicated equations like those used for the weather is likely to remain open for quite some time.

We cannot hope to defend a claim that a physical system is chaotic unless we discard the mathematicians' burden of proof, and with it the most common meaning of chaos. Nevertheless, if our best models of a physical system appear to be chaotic, if they are deterministic, appear to be recurrent, and suggest sensitive dependence by exhibiting the rapid growth of small uncertainties, then these facts provide a working definition of what it means for a physical system to be chaotic. We may one day find a better description of that physical system which does not have these properties, but that is the way of all science. In this sense, the weather is chaotic while the economy is not. Does this imply that if we were to add a so-called random number generator to our weather model we no longer believe real weather is chaotic? Not at all, as long as we only wish to employ a random number generator for engineering reasons, like accounting for defects in the finite computerized model. In a similar vein, the fact we cannot employ a true random number generator in our computer models does not imply we must consider the stock market deterministic. The study of chaos has laid bare the importance of distinguishing between our best models and the best way to construct computer simulations of those models. If our model structure is imperfect, our best models of a deterministic system might well turn out to be stochastic!

Perhaps the most interesting question of all to come out of chaotic forecasting is the open question of a fourth modelling paradigm: we see our best model fail to shadow, we suspect that there is no way to fix this model, either within the deterministic modelling scheme of our physicist, or within the standard stochastic schemes of our statistician. Can further study of mathematical chaos suggest a synthesis that will give us access to models that can at least shadow physical systems?

Shadows, chaos, and the future

> Our eyes once opened, we may pass on to a yet newer outlook of the
> world, but we can never go back to the old outlook.

A. Eddington (1927)

Mathematics is the ultimate science fiction. While mathematicians
can happily limit their activities to domains where all their
assumptions hold ('almost always'), physicists and statisticians
must deal with the external world through the data to hand and the
theories to mind. We must keep this difference in mind if we are
going to use words like 'chaos' when speaking with mathematicians
and scientists; a chaotic mathematical system is simply a different
beast than a physical system we call chaotic. Mathematics proves;
science struggles merely to describe. Failure to recognize this
distinction has injected needless acrimony into the discussion.
Neither side is 'winning' this argument, and as the previous
generation slowly leaves the field, it is interesting to observe some
members of the next generation adopt an ensemble approach:
neither selecting nor merging but literally adopting multiple
models *as a model* and using them in unison. Rather than playing
as adversaries in a contest, can our physicist, mathematician,
statistician, and philosopher work as a team?

The study of chaos helps us to see more clearly which questions
make sense and which are truly nonsensical: the study of chaotic
dynamics has forced us to accept that some of our goals are
unreachable given the awkward properties of nonlinear systems.
And given that our best models of the world are nonlinear – models
for the weather, the economy, epidemics, the brain, the Moore-
Spiegel circuit, even the Earth's climate system – this insight has
implications beyond science, extending to decision support and
policy making. Ideally, the insights of chaos and nonlinear
dynamics will come to the aid of the climate modeller, who, when
asked to answer a question she knows to be meaningless, is
empowered to explain the current limits to our knowledge and

communicate the available information. Even if model imperfections imply that there is no policy-relevant probability forecast, a better understanding of the underlying physical process has aided decision makers for ages.

All difficult decisions are made under uncertainty; understanding chaos has helped us to provide better decision support. Significant economic progress has already been made in the energy sector, where the profitability of using information-rich weather ensembles has led to daily use of uncertainty information from trading floors of the markets to the control rooms of national electricity grids.

Prophecy is difficult; it is never clear which context science will adopt next, but the fact that chaos has changed the goal posts may well be its most enduring impact on science. This message needs to be introduced earlier in education; the role of uncertainty and the rich variety of behaviour that mathematically simple systems reveal is still largely unappreciated. Observational uncertainty is inextricably melded with model error, forcing us to re-evaluate what counts as a good model. Our old goal to minimize least squares has been proven to mislead, but should we replace them with a search for shadows, for a model with good-looking behaviour, or the ability to make more accountable probability forecasts? From our new vantage point, we can see more clearly which questions make sense, calling forth challenges to the foundational assumptions of mathematical physics and to applications of probability theory. Are our modelling failures due to our inability to select the correct answer from among the available options, or is there no suitable option on offer? How do we interpret simulations from models which are not empirically adequate? Regardless of our personal beliefs on the existence of Truth, chaos has forced us to rethink what it means to approximate Nature.

The study of chaos has provided new tools: delay reconstructions that may yield consistent models even when we do not know the 'underlying equations', new statistics with which to describe dynamical systems quantitatively, new methods of forecasting uncertainty, and shadows that bridge the gaps between our models, our observations, and our noise. It has moved the focus from correlation to information, from accuracy to accountability, from artificially minimizing arguably irrelevant error to increasing utility. It rekindles debate on the status of objective probability: can we ever construct an operationally useful probability forecast, or are we forced to develop novel *ad hoc* methods for using probabilistic information without probability forecasts? Are we quantifying our uncertainty in the future of the real world or exploring the diversity in our models? Science seeks its own inadequacy; coping with constant uncertainty in science is not a weakness but a strength. Chaos has provided much new cloth for our study of the world, without providing any perfect models or ultimate solutions. Science is a patchwork, and some of the seams admit draughts.

Early in the film *The Matrix*, Morpheus echoes the words of Eddington that open this last section:

> This is your last chance. After this, there is no going back. You take the blue pill and the story ends. You wake up in your bed and you believe whatever you want to believe. You take the red pill and you stay in Wonderland and I show you how deep the rabbit hole goes. Remember that all I am offering is the truth. Nothing more.

Chaos is the red pill.

Glossary

> Mathematicians are like a certain type of Frenchman; when you talk
> to them they translate it into their own language, and then it soon
> turns into something completely different.
>
> Goethe, *Maxims and Reflections* (1779)

These entries are not meant to provide precise definitions, but
are intended to convey the main idea for quick reference. Some
terms hold different shades of meaning when used by
mathematicians (M), physicists (P), computer scientists (C), or
statisticians (S). Definitions and discussion can be found in the
CATS' Forum at *www.lsecats.org* and in books listed in the further
reading.

almost every (**M**): A mathematical catch phrase to warn that even
   though something is 100% true, there are instances when it is false.
almost every (**P**): Almost every.
attractor: A point or collection of points in *state space* which some
   other collection of states approach nearer and nearer as they are
   iterated forward.
basin of attraction: For a particular *attractor*, the collection of all
   states that will eventually approach it.
Burns effect: An expression that encapsulates the difficulty that
   incomplete foresight and imperfect models bring to attempts at
   rational decision making.

butterfly effect: An expression that encapsulates the idea that small differences in the present can result in large differences in the future.

chaos (C): A computer program that aspires to represent a chaotic mathematical system. In practice, all digital computerized dynamical systems are on or evolving towards a periodic loop.

chaos (M): A mathematical dynamical system which (a) is deterministic, (b) is recurrent, and (c) has sensitive dependence on initial state.

chaos (P): A physical system that we currently believe would be best modelled by a chaotic mathematical system.

chaotic attractor: An attractor on which the dynamics are chaotic. A chaotic attractor may have a *fractal* geometry or it may not; so there are *strange* chaotic attractors and chaotic attractors that are not strange.

conservative dynamical systems: A dynamical system in which a volume of *state space* does not shrink as it is iterated forward. These systems cannot have *attractors*.

delay reconstruction: A *model state space* constructed by taking time-delayed values of the same variable in place of observations of additional state variables.

deterministic dynamics: A dynamical system that can be iterated without recourse to a random number generator, whose initial state defines all future states under iteration.

dissipative dynamical system: A dynamical system for which, on average, a volume of *state space* shrinks when iterated forward under the system. While the volume will tend to zero, it need not shrink to a point and may approach a quite complicated *attractor*.

doubling time: The time it takes an initial uncertainty to increase by a factor of two. The average doubling time is a measure of predictability.

effectively exponential growth: Growth in time which, when averaged into the infinite future, will appear to be exponential-on-average, but which may grow rather slowly, or even shrink, for long periods of time.

ensemble forecast: A forecast based on the iterates of a number of different initial states forward (perhaps with different parameter values, or even different models) and in so doing reveals the diversity of our model(s) and so provides a lower bound on the likely impacts of uncertainty in model-based forecasts.

exponential growth: Growth where the rate of increase in X is proportional to the value of X, so that as X gets larger, it grows even faster.

fixed point: A state of a dynamical system which stays put; a stationary point whose future value under the system is its current value.

flow: A dynamical system in which time is continuous.

fractal: A self-similar collection of points or an object that is self-similar in an interesting way (more interesting than, say, a smooth line or plane). Usually, one requires a fractal set to have zero volume in the space that it lives, as a line in two dimensions has no area, or a surface in three dimensions has no volume.

geometric average: The result of multiplying N numbers together and then taking the Nth root of the product.

indistinguishable state: One member of the collection of points which, given an observational *noise* model, you would not expect to be able to rule out as having generated the observations actually generated by some target trajectory X. This collection is called the set of indistinguishable states of X and has nothing to do with any particular set of observations.

infinitesimal: A quantity smaller than any number you can name, but strictly greater than zero.

iterate: To apply the rule defining a dynamical *map* once, moving the state forward one step.

linear dynamical system: A dynamical system in which sums of solutions are also solutions, more generally one that allows superposition of solutions. (For technical reasons, we do not wish to say 'involves only linear rules'.)

Lyapunov exponent: A measure of the average speed with which *infinitesimally* close states separate. It is called an exponent, since it is the logarithm of the average rate, which makes it easy to distinguish exponential-on-average growth (greater than zero) from exponential-on-average shrinking (negative). Note that slower-than-exponential growth, slower-than-exponential shrinking, and no-growth-at-all are all combined into one value (zero).

Lyapunov time: One divided by the *Lyapunov exponent*, this number has little to do with the predictability of anything except in the most simplistic chaotic systems.

map: A rule that determines a new state from the current state; in this kind of mathematical dynamical system, time takes on only discrete (integer) values, so the series of values of X are labelled as X_i where i is often called 'time'.

model: A mathematical dynamical system of interest either due to its own dynamics or the fact that its dynamics are reminiscent of those of a physical system.

noise (measurement): Observational uncertainty, the idea that there is a 'True' value we are trying to measure, and repeated attempts provide numbers that are close to it but not exact. Noise is what we blame for the inaccuracy of our measurements.

noise (dynamic): Anything that interferes with the system, changing its future behaviour from that of the deterministic part of the model.

noise model: A mathematical model of noise used in the attempt to account for whatever is cast as real noise.

non-constructive proof: A mathematical proof that establishes that something exists without telling us how to find it.

nonlinear: Everything that is not linear.

observational uncertainty: Measurement error, uncertainties due to the inexactness of any observation of the state of the system.

pandemonium: *Transient dynamics* that display characteristics suggestive of chaos, but only over a finite duration of time (and so not recurrent).

parameters: Quantities in our models that represent and define certain characteristics of the system modelled; parameters are generally held fixed as the model state evolves.

Perfect Model Scenario (PMS): A useful mathematical sleight-of-hand in which we use the model in hand to generate the data, and then pretend to forget that we have done so and analyse the 'data' using our model and tools. More generally, perhaps, any situation in which we have a perfect model of the mathematical structure of the system we are studying.

periodic loop: A series of states in a deterministic system which closes upon itself: the first state following from the last, which will repeat over and over forever. A periodic orbit or limit cycle.

Poincaré section: The cross-section of a *flow*, recording the value of all variables when one variable happens to take on a particular value. Developed by Poincaré to allow him to turn a flow into a *map*.

predictability (M): Property that allows construction of a useful forecast distribution that differs from random draws from the final (climatological) distribution; for systems with attractors, this implies a forecast better than picking points blindly from the attractor.

predictability (P): Property that allows current information to yield useful information about the future state of a system.

prediction: A statement about the future state of a system.

probabilistic: Everything that is not unequivocal, statements that admit uncertainty.

random dynamics: Dynamics such that the future state is not determined by the current state. Also called stochastic dynamics.

recurrent trajectory: A trajectory which will eventually return very close to its current state.

sample-statistic (S): A statistic (for example: the mean, the variance, the average *doubling time*, or largest *Lyapunov exponent*) that is estimated from a sample of data. The phrase is used to avoid confusion with the true value of the statistic.

sensitive dependence (P): The rapid, exponential-on-average, separation of nearby states with time.

shadowing (M): A relationship between two perfectly known models with slightly different dynamics, where one can prove that one of the models will have some trajectory that stays near a given trajectory of the other model.

shadowing (P): A dynamical system is said to 'shadow' a set of observations when it can produce a trajectory that might well have given rise to those observations given the expected observational *noise*; a shadow is a trajectory that is consistent with both the noise model and the observations.

state: A point in *state space* that completely defines the current condition of that system.

state space: The space in which each point completely specifies the state, or condition, of a dynamical system.

stochastic dynamics: See *random dynamics*.

strange attractor: An *attractor* with *fractal* structure. A strange attractor may be chaotic or non-chaotic.

time series (M, P, S): A series of observations taken to represent the evolution of a system over time; the location of the nine planets, the number of sunspots, and the population of mice are examples. Also, the output of a mathematical model. Also (**S**): Confusingly, the model itself.

transient dynamics: Ephemeral behaviour as in one game of roulette, or one ball in either the Galton Board or the NAG Board, since eventually the ball stops. See *pandemonium*.

Further reading

For children

Michael Coleman and Gwyneth Williamson, *One, Two, Three, Oops!* (London: Little Tiger Press, 1999)

Fiction

Ray Bradbury, 'A Sound Like Thunder' (*Collier's Magazine*, 28 June 1952)

Carol Shields, *Unless* (Toronto: Random House Canada, 2002)

History of and historical science

Thomas Bass, *The Newtonian Casino* (Harmondsworth: Penguin, 1991)

Leon Brillouin, *Scientific Uncertainty and Information* (New York: Academic Press, 1964)

John L. Casti, *Searching for Certainty* (New York: William Morrow, 1991)

Arthur Eddington, *The Nature of the Physical World* (Cambridge: Cambridge University Press, Gifford Lectures Series, 1928)

E. E. Fournier d'Albe, *Two New Worlds* (London: Longmans Green, 1907)

Francis Galton, *Natural Inheritance* (London: Macmillan, 1889)

Stephen M. Stigler (2002) *Statistics on the Table: The History of Statistical Concepts and Methods* (Cambridge, Mass: Harvard University Press, 2002)

H. S. Thayer, *Newton's Philosophy of Nature* (New York: Hafner, 1953)

Philosophy of science

R. C. Bishop, *Introduction to the Philosophy of Social Science* (London: Continuum, in press)

N. Cartwright, *How the Laws of Physics Lie* (Oxford: Oxford University Press, 1983)

John Earman, *A Primer on Determinism* (Dordrecht: Reidel, 1986)

Jennifer Hecht, *Doubt: A History* (San Francisco: Harper, 2003)

P. Smith, *Explaining Chaos* (Cambridge: Cambridge University Press, 1998)

Chaos

L. Glass and M. Mackey, *From Clocks to Chaos* (Princeton: Princeton University Press, 1988)

Ed Lorenz, *The Essence of Chaos* (London: UCL Press, 1993)

J. C. Sprott, *Chaos and Time-Series Analysis* (Oxford: Oxford University Press, 2003)

I. Stewart, *Does God Play Dice?* (Harmondsworth: Penguin, 1997)

Weather

T. Palmer and R. Hagedorn, *Predictability* (Cambridge: Cambridge University Press, 2006)

More detailed discussions

Edward Ott, *Chaos in Dynamical Systems* (Cambridge: Cambridge University Press, 2002)

G. Gouesbet, S. Meunier-Guttin-Cluzel, and O. Menard (eds), *Chaos and its Reconstruction* (NOVA, 2003). (See, in particular, Chapter 9 by Kevin Judd for a review of ten years of work at CADO in dynamical systems modelling from time series.)

H. Kantz and T. Schreiber, *Nonlinear Time Series Analysis*, 2nd edn. (Cambridge: Cambridge University Press, 2003)

More on the Bakers, including the equations, can be found in H. Tong (ed.), 'Chaos and Forecasting', *World Scientific Publications* (Singapore, 1995).

The full 51-member forecast, along with a number of colour illustrations in this Very Short Introduction, can be found in L. A. Smith (2002) 'Predictability and Chaos', in *Encyclopedia of Atmospheric Sciences*, ed. J. Holton, J. Pyle, and J. Curry (New York: Academic Press, 2002), pp. 1777–85.

Also in this series

Timothy Gowers, Very Short Introduction to Mathematics

Index

Chaos

Chaos